ACID RAIN
Rhetoric and reality

ACID RAIN

Rhetoric and reality

Chris C. Park

METHUEN

London and New York

For Angela, with love

First published in 1987 by
Methuen & Co. Ltd
11 New Fetter Lane, London EC4P 4EE

Published in the USA by
Methuen & Co.
in association with Methuen, Inc.
29 West 35th Street,
New York NY 10001

© 1987 Chris C. Park

Printed in Great Britain by
Richard Clay Ltd,
Bungay, Suffolk

British Library Cataloguing in Publication Data

Park, Chris C.
Acid rain: rhetoric and reality.
1. Acid rain——Environmental aspects
I. Title
363.7'386   TD196.A2

ISBN 0-416-92190-6
ISBN 0-416-92200-7 Pbk

Library of Congress Cataloging in Publication Data

Park, Chris C.
Acid rain.

Bibliography: p.
Includes index.
1. Acid rain——Environmental aspects.
I. Title.
TD196.A25P36   1987   363.7'386
87-20398

ISBN 0-416-92190-6
ISBN 0-416-92200-7 (pbk.)

CONTENTS

Part IV THE POLITICS OF ACID RAIN

TABLES

FIGURES

ACKNOWLEDGEMENTS

Anyone who has ever written a book will know that when it is finished, the author swears *never* to write another one! This is my fifth, and I would not have tackled it without the support and encouragement of people who mean a great deal to me. Some of my partners in crime in the Department of Geography at Lancaster University helped directly. Peter Mingins and Claire Jarvis of the Cartographic Unit drew the figures; their skill in turning messy doodles into minor works of art is a constant source of pleasure. Clair Gould, formerly of the Physical Geography Laboratory, clarified some of the science where my understanding was minimal, and helped to check some of the references; her quiet enthusiasm is much appreciated. My fellow academics in the department, as ever busy with their own research, kept out of my way and provided a pleasantly quiet environment in which to work.

Writing is a solitary pursuit, and my sanity and vanity were kept in check by my cherished circle of friends. Claire Jarvis, Philip and Gill Gower, Peter-John and Ann Davies, and Robin and Jill Bundy helped in various ways to take my mind off acid rain and my hands off the word-processor; I really value their friendship. Angela, my best friend and wife, witnessed my alternating phases of agony and ecstasy in writing the book, with customary calmness. Her disarming indifference to things I find absorbing helps me to keep things in perspective. This book is an inadequate repayment for her honesty, love, and support.

PREFACE

Acid rain is now one of the most serious environmental problems in developed countries. Interest in the problem has increased significantly within the last decade, especially since reports of extensive forest dieback in West Germany (even in areas considered to be unpolluted) started to appear in 1975. Concern quickly spread to Scandinavia, where large-scale fish deaths were believed to be caused by acid rain, and where more recently concern has focused on forest changes and heavy metal mobilization in streams, lakes, and groundwater.

Acid rain is widely believed to result from the washout from the atmosphere of oxides of sulphur and nitrogen. Although these oxides exist naturally in environmental cycles and ultimately have natural sources, the main sources today are coal-fired power stations and smelters (which produce sulphur dioxide, SO_2) and motor vehicle exhausts (which produce nitrogen oxides, NO_x). These gaseous pollutants are light and invisible, and they can be carried hundreds if not thousands of kilometres by prevailing winds. The oxides may then mix with other chemicals in the atmosphere to produce the poisonous and corrosive substances that either settle as dry fallout or are washed out by rain as acid deposition.

Emissions and fallout were previously extremely localized, but since the introduction of 'tall stacks' policies in both Britain (since 1958) and the United States (since 1971) – paradoxically to disperse particulate pollutants and hence reduce local damage – emissions are now lifted into the upper air currents and carried long distances downwind. The tall stacks policies have thus turned a smoke problem into an acid rain problem.

Neither winds nor acids respect political boundaries, so the pollutants are often carried across state and national frontiers. Acid rain thus represents a hidden export from one country to another – the exporter gains through not having to install costly pollution-control equipment or constrain industrial activity to limit emissions, and the importer suffers from the adverse impacts on its environment (with none of the economic gains from the pollution that are enjoyed by the exporter, whose air is purified in the process). Thus acid rain is a critical problem for international relations, and it rightly claims a high priority on the agenda of political debate both within and between countries. It is as much a question of ethics as of practice.

The acid rain debate now embraces many western countries – including Canada, the United States, England, Scotland, Wales, Sweden, Norway, Denmark, West Germany, the Netherlands, Austria, Switzerland – and a growing number of eastern countries – including the Soviet Union, Poland, East Germany, Czechoslovakia. Acid rain is believed to blow over Scandinavia and West Germany from all over Europe, but Britain is often singled out as the largest single culprit (after the Soviet Union). The United Nations Environmental Programme views acid rain as 'a particularly modern, post-industrial form of ruination [which] is as widespread and careless of its victims and of international boundaries as the wind that disperses it'.

The problem of acid rain arises, strictly speaking, not so much from the rainfall itself as from its effects on the environment. Runoff affects surface water (such as rivers and lakes) and groundwater, as well as soils and vegetation. Consequently changes in rainfall acidity can trigger off a range of impacts on the chemistry and ecology of lakes and rivers (over 200 lakes in the Adirondack Mountains of New York state are devoid of fish), soil chemistry and processes (e.g. the increased mobility of toxic and persistent heavy metals such as cadmium), the health and productivity of plants (including trees and crops; e.g. it is estimated that around one-third of West Germany's forests are now seriously affected by acid rain), and building materials (including those of historic and cultural importance, such as many of England's finest cathedrals), and metallic structures (such as New York's Statue of Liberty). Human health might also be affected via the intake of food and water contaminated with toxic heavy metals (e.g. cadmium) mobilized by fallout of acid rain. The significance of such impacts is more than purely scientific – they affect the quality of life for humans, they threaten environmental stability and the sustainability of food and timber reserves, and they pose real economic problems (e.g. in the cost of repair to buildings). Acid rain is thus very much a human as well as a scientific problem, and it has broad economic, social, and medical implications.

The most suitable solutions to the problems of acid rain require prevention rather than cure, and there is broad agreement in both the political and scientific communities on the need to reduce emissions of sulphur and nitrogen oxides to the atmosphere. The technology now exists to make this possible. However, clean-up programmes will be expensive, and they might yield only qualified improvements. Inevitably the high costs involved must be met by the countries that export the pollution, whereas the benefits to be gained will be enjoyed most by the importer countries. Thus arises the inevitable difference of opinion between countries over the best way of tackling the problem of acid rain.

Those countries that are hardest hit by the problem (such as Canada and the Nordic states) are convinced that something must be done, without further delay, to reduce emissions and thus save their forests and lakes. With the belief that 'charity begins at home', some twenty-one governments have now resolved to cut their own emissions of sulphur dioxide by at least 30 per cent (over 1980 levels) by 1993, and hope to persuade the main producers of the culprit oxides to join them in a truly international attack on the problem. The real villains of the piece, however, are the United States and Great Britain, who stubbornly refuse to join this '30 per cent club' and remain resolute in their belief that the problem is not yet sufficiently well understood to warrant expensive investments which might at best yield only minor improvements in environmental quality.

In December 1984 the UK Parliamentary Select Committee on the Environment reported on acid rain. It concluded that Britain is the worst air polluter in Western Europe, and that it should take urgent action to reduce emissions of sulphur and nitrogen oxides (mainly from coal-fired power stations). The British government, however, presently accepts the proposed European emission limits as targets to aim for rather than obligations to fulfil.

The international scientific community regards acid rain as one of the most serious and significant environmental problems of our time, which requires urgent attention and immediate political initiatives both nationally and internationally. Leading politicians and government officials in both Britain and the United States argue that the problems are as yet too little understood, and that links with possible source areas and likely sources (such as coal-fired power stations) are as yet too ill defined to justify wholesale political intervention, which would require the introduction of emission controls and lead to electricity price rises for the consumer. A central part of the acid rain debate thus centres around the balance between scientific and political arguments. Of particular interest are the reasons given for reluctance of the British government to co-operate with international initiatives designed to reduce emissions and eliminate acid rain. This attitude is unpopular among other member states of the European Community, antagonistic to the initiatives being formulated by the Scandinavian countries, and – according to many experts – curiously at odds with available evidence. The acid rain debate is thus as much a battle in politics as a struggle in science.

Part I

THE PROBLEM
OF ACID RAIN

1 THE ACID RAIN DEBATE IN CONTEXT

'Long to reign over us . . .'

> this most excellent canopy, the air, look you, this brave o'erhanging firmament, this majestical roof fretted with golden fire, why, it appeareth nothing to me but a foul and pestilent congregation of vapours.
>
> (William Shakespeare, *Hamlet*, II, ii, 299)

Acid rain has been called 'an unseen plague of the industrial age' (Anon 1984b: 6), and it is generally regarded as one of the most serious environmental problems of our times. As an issue with both scientific and political dimensions it ranks alongside important contemporary concerns like the global increase of carbon dioxide in the atmosphere, the spread of toxic chemicals in the environment, and the possible environmental consequences of nuclear war. However, it presents a somewhat unique problem in that its consequences are already evident, its adverse effects are already documented, and its impacts are very real to people living in affected areas (Whelpdale 1983: 72).

Acid rain is a widespread, serious, and costly problem that will not simply fade away in the public consciousness or cease to create environmental damage. It has been blamed for damage to trees in West Germany, death of fish in Scandinavia, and other unwanted effects (including damage to buildings and human health), and it threatens international relations between the United States and Canada, and between Britain and her European neighbours. The Organization for Economic Co-operation and Development (OECD) has estimated that acid rain costs the countries of the European Community £33–44 billion per annum (*The Times*, 19 November 1985, p. 16).

This book seeks to review the main arguments currently being levelled in the 'acid rain debate'. We shall examine the general context of acid rain in this chapter, and then progress to examine the effects it is believed to have on lakes and rivers, forests and crops, humans and buildings (in Part II) and to explore possible solutions and remedies (in Part III). The diplomatic and political ingredients of the debate are considered in Part IV.

The term 'acid rain' refers to the dilute sulphuric and nitric acids which, many believe, are created when fossil fuels are burned in power stations, smelters, and motor vehicles, and which fall over areas long distances downwind of possible sources of the pollutants. For convenience the term 'acid rain' will be used throughout this book, although it has been dismissed as 'emotive' and 'inadequately named' (correspondence by Professor B.A. Thrush to *The Times*, 22 September 1984, p. 9). The term is a misleading short-hand version of 'acid precipitation', which includes the dry fallout of oxides of sulphur and nitrogen (in the form of dry gases and minute aerosols, particles that remain suspended in the atmosphere) as well as wet deposition of acids (in solution or suspension in fog, or on raindrops, snowflakes, or hail). 'Acid rain', as used here, includes both dry and wet deposition.

ACID RAIN AS A POLLUTANT

Pollutants can be classified into two groups (Park 1981: chapter 7). First, there are man-made materials (such as persistent synthetic chemicals like DDT), which are not part of natural environmental cycles and therefore do not readily break down when released into the environment. A sub-set of this group would include materials like nuclear waste products that are – strictly speaking – parts of natural cycles, but these cycles operate extremely slowly over long time-spans. The second group comprises materials that already exist naturally in the environment, and that natural environmental processes and cycles can cope with (or neutralize), break down, disperse and recycle, but that appear in much higher concentrations than would normally be the case. Pollutants of this type are not necessarily harmful in themselves – they create problems only when they overload natural biogeochemical cycles.

Acid rain falls in this second group because its basic ingredients (sulphur dioxide, SO_2, nitrogen oxides, NO_x, and ozone, O_3) do appear naturally in the environment, albeit in smaller concentrations. In fact the natural acidity of rainfall provides free supplies of valuable nutrients for plant growth. Some areas (even parts of Scandinavia) with mineral-deficient soils are happy to receive the sulphur because otherwise costly artificial fertilizers would be required. It is the increased acidity experienced in recent years (which many associate with air pollution) that appears to produce serious environmental problems, and which is the true focus of the acid rain debate.

As a form of pollution, acid rain has some unusual properties. It is invisible, with no discernible taste or smell to humans. It has remained largely undetected even in areas where it has been falling for many years, because its effects on the environment are not readily noticeable in their early stages. It has no rapid, dramatic effects; it is a silent, creeping paralysis form of cumulative pollutant. Only in advanced cases are its effects sufficiently quantifiable to be convincing in any statistical sense (Elder 1983: 58).

Neither does acidification normally cause unsightly effects on the landscape, of the sort often associated with nutrient enrichment (eutrophication) with nitrate fertilizers washed from fields, where enriched lakes are draped with dense floating mats of algal blooms. In fact, few who have witnessed Scandinavia's acidic lakes fail to be moved by the air of tranquillity offered by the clear lakes, devoid of smell or floating vegetation, with visibility increased considerably (through loss of aquatic life), and lake beds covered by white acid-tolerant mosses. Acidified lakes are reminiscent of Rachel Carson's *Silent Spring* (1962), in which birds and animals died from the effects of toxic pesticides and insecticides like DDT.

THE GEOGRAPHY OF ACID RAIN

Acid rains and the oxides that create them are blown long distances by the wind, often crossing seas and national frontiers to become an invisible export. As a result, the polluted is often far removed downwind from the polluter, who may remain unconvinced that a real problem exists. The polluter may then raise serious and well-intentioned objections to any remedial measures proposed by the polluted that would impose a heavy cost on itself.

Added complexity arises from the fact that, although sulphur and nitrogen oxides are now relatively ubiquitous in industrial nations, not all areas are affected by acid rain, and different locations (even within an area) show different symptoms of acidification. As yet, the areas where acid rain impacts are noticeable have been relatively few and predictable, given the ingredients of the 'acid rain equation'. The most seriously affected areas are shown in Figure 1.1. They tend to have a number of properties in common (La Bastille 1981: 660):

- they are concentrated in the industrialized belt of the northern hemisphere, downwind of dense concentrations of power stations, smelters, and large cities;
- they are often upland or mountainous areas, which are well watered by rain and snow;
- being well watered, they are often dissected by lakes and streams, and often covered by forest; and
- being upland, they often have thin soils and glaciated bedrock.

Many parts of Scandinavia, Canada and the Northeast United States, and northern Europe (particularly West Germany and upland Britain) share these properties, and this is why they figure so prominently in the acid rain debate. Across the Atlantic there are a number of 'acid rain hot spots', including Nova Scotia, the Canadian Shield around southern Ontario and Quebec, the Adirondack Mountains, Great Smoky Mountains, parts of Wisconsin and Minnesota, the Pacific Northwest USA, the Colorado Rockies, and the Pine Barrens of New Jersey (La Bastille 1981).

In contrast, there are two types of 'safe area', where acid rain is not a

Figure 1.1 The global nature of acid rain in the early 1980s

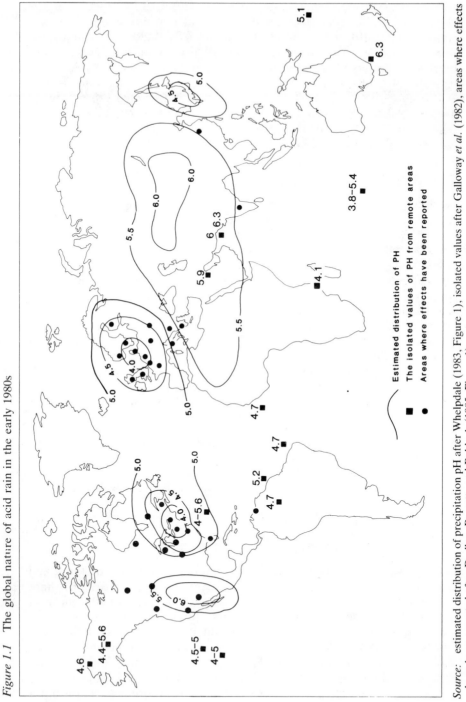

Source: estimated distribution of precipitation pH after Whelpdale (1983, Figure 1), isolated values after Galloway *et al.* (1982), areas where effects have been reported after Dudley, Barrett, and Baldock (1985, Figure 1).

problem (at present). One comprises those areas that simply do not receive acid rain or the gaseous oxides of sulphur and nitrogen, because of the fortunes of location (away from and not downwind of possible source areas). Almost all of the southern hemisphere, the tropics and parts of the northern hemisphere are so protected (up to now). The second group comprises areas that receive acid precipitation (wet or dry) but can tolerate it. Many areas have a natural resistance (or 'buffering capacity' – see chapter 3) to acidification, with immunity offered by alkaline soils or limestone bedrock, which neutralize acid inputs. Other forms of buffering are offered in the Midwest United States, where alkaline dust blown from the west neutralizes acid rain before it reaches the ground (La Bastille 1981). Alkaline precipitation has been observed in Sweden in the past (pre-1960) in areas with limestone outcrops and areas where cement was manufactured (Barrett and Brodin 1955).

ACID RAIN IN HISTORIC CONTEXT

Concern over acid rain may be a product of the last decade, but acid rain itself most certainly is *not*. It is very much a product of the industrial age. Although a natural background level of acid rain (derived from natural sources such as volcanic eruptions) has existed through the ages, it is only within the last 200 years that the widespread burning of fossil fuels has led to a dramatic (man-made) increase in emissions of sulphur and nitrogen oxides, and a corresponding increase in rainfall acidity.

There is evidence from many areas of this association in time. Some of the evidence reflects ecological changes triggered by acid precipitation in the past (which are described in greater detail in chapters 4 and 5). For example, the bogmoss (*Sphagnum* spp.) is very susceptible to SO_2, and there are historic records of its disappearance from the Pennines at the time of the industrial revolution (Tallis 1964; Ferguson and Lee 1983). Records of changing diatom (microscopic marine or freshwater plants) populations, which reflect increasing acidity of the lake waters during the last two centuries, are preserved in lake sediments in some Scottish lochs (in Galloway, which presently receives acid rain) (Battarbee and Flower 1982; Flower and Battarbee 1983; Battarbee 1984; Flower 1984; Pennington 1984; Battarbee *et al.* 1985a, b). Many species of lichen are also intolerant of SO_2, and lichen deserts are common in and around urban and industrial areas in Britain and elsewhere (Gilbert 1975).

Other evidence comes from studies of changing patterns of rainwater chemistry. Direct evidence of a ten-fold increase in atmospheric nitrate concentrations in North America and a five-fold increase in Europe since the turn of the century emerges in analyses of records made by Victorian scientists and kept at agricultural stations in Europe and North America (Brimblecombe and Stedman 1982). Indirect evidence is preserved in the chemical impurities locked up in the unique archive of polar snow and ice. As each new snow layer is incorporated into the ice, material from the overlying air is trapped within it as both particles and gases (in air bubbles). The precise age of different levels within the ice can be determined scientifically, and so the polar ice caps contain an invaluable record of pollution spanning up to half a million years. Recent analyses of ice cores from Greenland (Wolff and Peel 1985; also *The Times*, 24 May 1986) indicate that atmospheric concentrations of sulphate and nitrate were variable but low before the turn of the century (reflecting background levels of natural origin), but since 1900 they have increased exponentially – nitrates have doubled and sulphates have trebled.

Apart from providing a long record of changing air quality, these Greenland studies highlight the widespread distribution of the oxides (which are dispersed by prevailing upper air winds). Recent studies of ice cores from the Antarctic (Wolff 1986), which confirm that there has been no detectable increase in sulphates and nitrates in that part of the southern hemisphere (remote from industry and emission sources of pollutants) over the last fifty years, provide an interesting and salutory contrast.

The evidence of association in time complements the evidence of association in space between observed patterns of acid rain and man-made emissions of SO_2 and NO_x. This does not in itself prove cause and effect, but the associations have clearly not arisen by chance.

HERITAGE OF INTEREST

Widespread scientific interest has only started to focus on acid rain within the last decade. This reflects the growing seriousness of the problem and the related rise of acid rain as a political issue, but it is slightly surprising given over a century of research by scientists. There are four leading figures in this intellectual dynasty (Cowling 1982; La Bastille 1981: 661–2).

Robert Angus Smith, an English chemist who was Britain's first Alkali Inspector (i.e. air pollution watchdog), was the subject's founding father. In 1852 Smith discovered a link between the sooty skies over industrial Manchester and the acidity he found in precipitation, and twenty years later he used the term 'acid rain' in a 600-page book on the subject (Smith 1872; Gorham 1982). Smith's ideas were largely ignored until they inspired more detailed research on rainwater quality and its control in the Lake District in north-west England during the late 1950s by Dr Eville Gorham, a Canadian ecologist (Gorham 1955, 1958).

By the mid-1960s early symptoms of acidification were starting to appear in Scandinavia, and in 1967 the Swedish soil scientist Svente Oden began a campaign to inform the international scientific community about acid rain, its assumed causes, and its possible effects by speaking on it at scientific meetings and writing about it in academic journals (Oden 1976). Oden – widely regarded as the father of acid rain studies – envisaged a new form of 'chemical warfare' in Europe involving the effects of acid rain on surface water chemistry, fish populations, forest growth, plant diseases, and accelerated damage to buildings. His campaign, pursued with missionary zeal, kindled an interest in acid rain among reputable scientists that, although slow to start, has accelerated since.

Studies in the United States date back to 1963 (at least a decade before acid rain was to become a major issue in North America), when Dr Gene Likens and Dr F. Herbert Bormann embarked on a multidisciplinary study of a small watershed within the Hubbard Brook Experimental Forest in New Hampshire, which was to continue at least up to the present time. Their study included the chemistry of rainwater, and they found that highly acidic rains fell over the area, although it was remote from possible local sources of oxides (Likens, Bormann and Johnson 1972; Likens *et al.* 1977).

Acid rain is not simply an item of scientific curiosity. It has become a major issue in international diplomacy. It was first raised as an international issue by Sweden at the United Nations Conference on the Human Environment in Stockholm in 1972, and through the next decade scientists and politicians in many countries started to realize the possible scale and significance of the problem. International dialogue about acid rain has increased considerably since 1980, and it has become a key item on the agenda of domestic and international political debate in Europe, Scandinavia, Canada, and North America. The so-called 'acid rain debate' is going on in Britain and the United States, and in many other developed countries. It will be examined in Part IV.

POLARIZATION OF VIEWS

There are two main schools of thought in the debate (Barabas 1983: 114). The majority of scientists, and many Canadian and European politicians, attribute visible damage to lakes, forests, crops, and buildings to man-made emissions of sulphur dioxide (SO_2) and nitrogen oxides (NO_x) from the burning of fossil fuels (mainly in conventional coal-fired power stations). In Britain the scientists and politicians are joined by a range of pressure groups (including Friends of the Earth, Greenpeace, the Ecology Party, the Socialist Environment and Resources Association, and the Young Liberals Ecology Group) who have banded together to forward the 'Stop Acid Rain UK' campaign. This unlikely and at times uncomfortable coalition demands the introduction of remedial measures without delay to significantly reduce emissions from chimneys and thus protect the environment from further damage.

A minority of scientists, and a caucus of influential politicians in Prime Minister Thatcher's Cabinet and President Reagan's administration, are *not* convinced that the case against man-made oxide emissions is proven or that costly emission reductions imposed on industry would bring about the desired results. These dissenters are therefore resolutely opposed to the adoption of emission control measures, and repeatedly call for more studies on the causes of acid rain.

The relentless exchange of acid rhetoric between these two schools has created a fascinating debate combining science, politics, and diplomacy in

Britain (see chapter 10) and elsewhere (see chapters 8 and 9). The debate has continued for over a decade and by the late 1980s there are still no signs on the horizon of any satisfactory (or even honorable) resolution. The debate ranges far and wide, although there is now little doubt that the acidity of rainfall in many areas has increased in recent years and continues to rise, and no doubt that man-made emissions of SO_2 and NO_x can – if necessary – be reduced (*at a cost*).

However, there remain significant differences of opinion on a number of key areas, including:

- the extent to which man-made emissions of SO_2 and NO_x cause acid rain (i.e. might not natural sources and processes play a significant role?);
- the extent to which observed damage to lakes and rivers, forests, crops, and buildings can be attributed definitely to acid rain (i.e. are the inferred links real and causal, or are other – perhaps as yet hidden or unexplored – factors also involved?);
- the extent to which observed damage is significant (i.e. is the damage serious enough to justify intervention to reduce if not reverse it, and if recovery is possible how long might it take?);
- the extent to which proposed reductions of oxide emissions would be effective (i.e. would there be a proportionate reduction in damage to the environment?); and
- the extent to which proposed reductions of oxide emissions would be cost-effective (i.e. would the value of the benefits – such as improved forestry and fishing yields and less tangible benefits like improved environmental quality – equal or exceed the costs of installing and running equipment to reduce emissions from chimneys?).

These differences of opinion highlight the complexity of the acid rain debate, because they all relate to relative questions to which there are no black and white answers. Consequently, because each side in the debate starts from different premises and is in pursuit of different goals (protection of environment versus protection of industry and free enterprise). It is extremely difficult for a compromise position to emerge. There is only a winner and a loser in the debate.

In the final analysis, the environment – on which we depend for natural resources (such as food, timber, and materials) and intangibles like quality of life and scenic landscapes – will be the winner or loser.

THE DEBATE OPENS

Britain has long claimed to lead the world in pollution control. The claim is not without foundation in a country that has pioneered the monitoring and control of air and water pollution, and that has a long and distinguished history of effective environmental legislation. But in the present international debate over acid rain, Britain has elected to play a very different role (outlined in detail in chapter 10), almost diametrical to its conventional trail-blazing stance.

While most European countries are pressing ahead with important initiatives designed to reduce emissions of sulphur dioxide and nitrogen oxides (see chapter 8) in an attempt to reduce if not eliminate acid rains from their skies, Britain chooses to 'go it alone' and has adopted an attitude of inertia and apparent lack of concern that has proved frustrating to its European partners and to many scientists at home and abroad. On the other side of the Atlantic, Canada is following the European lead in reducing emissions, while the United States plays the British game of maintaining a low profile and preferring to hope that the problem will soon become yesterday's news (see chapter 9).

Those responsible for protecting Britain's environment have not been slow to point to past successes in dealing with pollution. Mr Michael Heseltine, then Secretary of State for the Environment, told a United Nations Environment Conference held in London in June 1982:

> I believe it is fair to claim that genuine, proven air pollution problems have been less severe [in Britain] in the last ten years. I believe I can say with confidence that in the United Kingdom, in other countries in Europe, and I believe in North America, Japan and many other industrialized states, the rate of progress is faster than the rate of deterioration. I believe that we are winning.
>
> (*The Times*, 16 June 1982, p. 8)

Not everyone would share his optimism, and there is a growing belief that, although Britain has won some minor skirmishes, the war is far from won. Moreover, policies and actions taken to reduce earlier air pollution problems are not unrelated to present-day acid rain problems. In fact with the benefit of hindsight it appears that they have been prime catalysts.

Britain's grimy past

In the past, air pollution was an extremely localized problem, centred mainly on towns and cities. In the fourteenth century, for example, Edward I ordered a man to be tortured for burning coal and fouling London's air (Goudie 1981: 261). Problems of air quality have become much more acute and more widespread within the last 200 years, with rapid increases in the size of urban areas and the spread of industrialization. Britain was to lead the world in industrial development, but the toll for this was dark skies, grimy buildings, and an unhealthy urban population – all brought about by air pollution from the burning of fossil fuels.

In the nineteenth century, much of Britain's air was smoky and acrid. The nation's industrial cities were the dirtiest in the world, although most of the countryside escaped the worst excesses of grime and toxic gases. The conditions in Carlyle's 'sooty Manchester' (Briggs 1980: 93) are vividly captured in a contemporary account of the city given in 1835 by Alexis de

Tocqueville, a visiting French aristocrat, who saw a land 'given over to industry's use', in which 'huge palaces of industry . . . keep air and light out of the human habitations which they dominate; they envelope them in perpetual fog . . . a sort of black smoke covers the city. The sun seen through it is a disc without rays . . .' (Harvie, Martin, and Scharf 1981: 40–1).

Well into the present century, the traditional heavy industrial cities of Britain – workhorses in the nation's struggle to stay ahead of the world economic pack – paid the price for continued reliance on fossil fuels in terms of satanic environments. Ronald Kershaw describes Sheffield in the 1930s, where 'every square foot of exposed masonry was black and filthy. The sun rarely penetrated a vast umbrella of smoke and soot. Children were taken away from the city for the sake of their health. There were thousands upon thousands of back to back houses. Their occupants were as grim and grimy as the steelworks from which they drew their sustenance' (*The Times*, 19 December 1981, p. 22).

This was the depressed baseline against which Mr Heseltine drew his comparisons, and against which his opponents clearly are winning. There are several landmarks in Britain's historic and pioneering fight against air pollution. The Alkali Act of 1863 tackled the worst types of air pollution, beginning with hydrochloric acid gas emissions from the old alkali industry (Department of Environment 1984: 3). A major turning point was London's infamous 'Great Smog' of 5–9 December 1952, the best known and most serious of a long line of 'pea-soupers' that had afflicted the capital city since Dickensian times and been recorded for posterity in the Sherlock Holmes stories. Sulphur dioxide gas combined with sooty emissions from chimneys to form acid soot in the moist air. The smog led to an estimated 4,000 additional deaths from lung and heart diseases, mostly amongst the elderly and infirm, in the weeks and months that followed (Royal College of Physicians 1970).

Tighter laws and clearer skies

In the wake of the 1952 smog, the government commissioned the Beaver Committee to evaluate levels and determine the main sources of air pollution and recommend appropriate controls. The committee, whose report was published in 1954, estimated that air pollution was costing the nation over £300 million per year (at 1954 prices) and proposed the central ingredients of a pollution abatement policy, which were to be embodied in legislation through the 1956 and 1968 Clean Air Acts.

The Clean Air Acts focused mainly on visible air pollution, in the form of smoke emissions containing particulate material. The main aim was to cut soot emissions from chimneys and thus prevent the formation of smog particles. Invisible gaseous pollutants, such as sulphur dioxide, were almost completely overlooked in the Acts. The problem of smoke was tackled in various ways (Department of Environment 1984: 3). The greater part of Britain's smoke had been coming from domestic chimneys, and this source was reduced by enabling local authorities to declare smoke control zones, within which the emission of smoke from housing was banned, and to help people to pay for the installation of special grates for burning smokeless fuel.

Power stations and industry had been emitting less smoke than houses, and the favoured solution was to remove larger particulate material at source and build higher chimneys to release the remaining small particulates and gases higher in the atmosphere where diffusion and dispersion over a wider area would be possible. Smoke from lorries, buses, and other diesel-engined vehicles has been reduced by setting tighter standards to be met by manufacturers, and by enforcing better mainte-nance through annual inspections. Smoke from the railways has been all but eliminated with the demise of the 'steam age' attendant upon the switch from coal-burning to electric and diesel power. It is important to stress that these air pollution control measures were not designed to stop or limit the production of sulphur dioxide and nitrogen oxides; nor did they. The aim was to disperse the gaseous emissions over a wider area (thus diffuse and dilute them).

These pollution-control measures, coupled with the switch from coal to cleaner, more convenient fuels (such as natural gas and oil), have led to a fall in smoke emissions in the United Kingdom by over 85 per cent since 1958 (Figure 1.2) and brought visible improvements in air quality over Britain. The 'tale of three cities' is now very different. Ronald Kershaw claims with pride that Sheffield is now smoke free and 'one of the cleanest industrial cities in Europe' (*The Times*, 19 December 1981, p. 22). Cleaner air means more sunshine, and the decline in smoke concentrations over central London since 1960 (Figure 1.3) has brought an increase of about 70 per cent in the amount of winter sunshine recorded at the London Weather Centre in central London (Figure 1.4). On an average winter day, visibility in London has more than doubled. Londoners are thankful that thick, choking smogs are now a thing of the past. Manchester boasts a 90 per cent reduction of winter smoke, a doubling of winter sunshine hours, a fall in sulphur dioxide levels by two-thirds, and a halving of the bronchitis death rate since the mid-1950s (*The Times*, 18 March 1980, p. 4). Oxford has also enjoyed a marked drop in the frequency of fogs since the mid-1960s, attributed in part to implementation of the 1956 Clean Air Act (Gomez and Smith 1984).

Improvements have been seen in the invisible pollutants as well. Sulphur dioxide emissions across the nation have fallen by one-third since 1970 (Figure 1.5), and by up to a half in many cities (including London) (Department of Environment 1984), mainly because of declining outputs from domestic, transport, and industrial sources. Clearly not all of the improvement in air quality can be attributed directly to the effect of legislation, because of the delay in improvement evident in many areas;

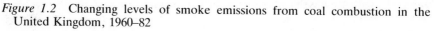

Figure 1.2 Changing levels of smoke emissions from coal combustion in the United Kingdom, 1960–82

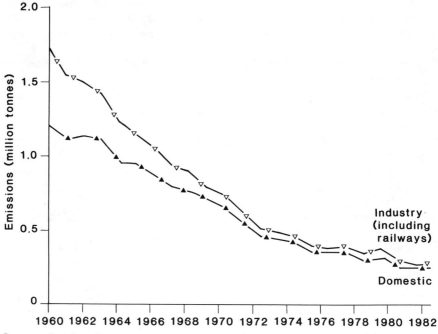

Source: data from Department of Environment (1984)

Figure 1.3 Sulphur dioxide and smoke concentrations (annual averages) in central London, 1962–83, and estimated sulphur dioxide concentrations over the whole of London since 1580

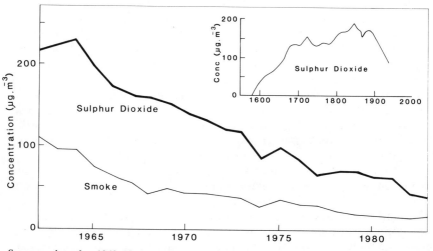

Source: data for 1962–83 from Greater London Council (1985); data since 1580 from Brimblecombe (1977)

Figure 1.4   Increasing levels of winter sunshine in the London area, 1950–82

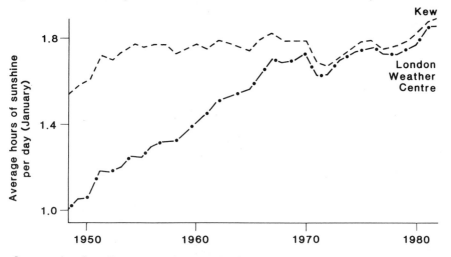

Source: data from Department of Environment (1984)

social as well as legislative factors are involved, along with the changing industrial base and closure or conversion of public utilities.

There are many signs of improving air quality in post-war Britain, and in 1972 Mr Eldon Griffiths (Under-Secretary of State at the Department of the Environment) could proudly claim that 'Britain leads the industrial world in cleaning up its urban air' (*The Times*, 21 April 1972). Kenneth Mellanby has noted how 'in the last 30 years our cities have all become much cleaner, human health has improved, plants flourish where previously they died, and damage to our buildings has been greatly reduced' (correspondence to *The Times*, 15 September 1984, p. 9). Oak trees, which are intolerant of smoke and industrial fumes and had almost disappeared from London since the industrial revolution, could be planted again in Hyde Park in 1978 (*The Times*, 6 March 1978, p. 2). Even through the 1970s there was mounting evidence of declining air pollution in London, with the gradual return of lichen flora in woodlands in the north and west of Greater London, including some species last seen there in 1800 (Hawksworth and Rose 1981), possibly reflecting the gradual decline of sulphur dioxide concentrations over the nation's capital city (Figure 1.3).

LEGISLATION – SUCCESS OR FAILURE?

In the light of this apparent success story of controlling air pollution and improving Britain's air quality, it is interesting to reflect on why attitudes towards air pollution control in Britain have altered so visibly, and to ask what is the connection with acid rain? Two key factors are intertwined.

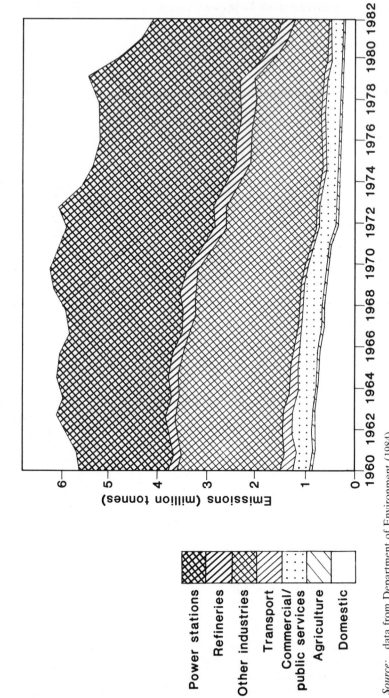

Figure 1.5 Sulphur dioxide emissions from fuel combustion in the United Kingdom, 1960–82

Source: data from Department of Environment (1984)

Visible smoke and invisible gases

The first key factor is mounting concern over the invisible (i.e. gaseous) air pollutants, particularly sulphur dioxide (SO_2). Despite evidence of a national fall in emissions (see Figure 1.5), and of improved air quality at many city sites, there was growing anxiety amongst informed observers during the 1970s that concentrations of SO_2 from power stations and home heating systems, and carbon monoxide from cars, were well above the limits in force in some countries (Greater London Council 1973).

The anxiety was fuelled partly by emerging evidence of damage that appeared to stem from air pollution. For example, some well-established plant species that had flourished for nearly 300 years in London's Chelsea Physic Garden (one of the oldest botanical gardens in the world) were showing signs of stress by 1976. The possible links between these observed effects and measured levels of SO_2 in the Garden (which proved to be three times higher than in industrial Teesside in north-east England) did not pass unnoticed (*The Times*, 25 March 1976).

In December 1973 Dr Basil Brown, Scientific Adviser to the Greater London Council (GLC), had warned that London – which already had amongst the highest SO_2 levels in the country – faced the threat of increased pollution from SO_2 emitted from new power stations planned for the capital over the next decade (Greater London Council 1973). Almost a decade later, in August 1982, the Scientific Branch of the GLC was to report that central London, which still suffered more SO_2 pollution than any city in Britain, was 'nudging the limits set down by the EEC' (Greater London Council 1982).

Manchester's Assistant Director of Environmental Health expressed fears in March 1980 that the improvements in SO_2 levels observed in recent years were unlikely to continue (in fact they might well deteriorate) over the next decade, particularly in the city centre. His predictions were founded in the increasing tendency amongst both industrialists and householders to return to coal as a source of heating and energy in a financial climate of restraint and in the light of OPEC oil price increases (which were forcing oil users to turn to heavier grades with a higher sulphur content, or to switch to other fuels).

Tall stacks paradox

The second critical factor, not unrelated to the first, was the tall chimneys (or stacks) policy introduced in the Clean Air Acts. The new taller chimneys which discharge pollution high in the air, did improve air quality locally by dispersing the pollutants over a wider area. An 'out of sight, out of mind' attitude to air pollution control prevailed. Dilution was the solution to pollution, and the 'air pollution problem' of the day, defined on the basis of particulate material, appeared to be solved.

However, the 'problem' was not solved – particularly for gaseous pollutants – because what goes up must come down, even if it is further afield. The sulphur and nitrogen oxides released on burning fossil fuels can be carried hundreds if not thousands of kilometres by the winds, and then fall (as wet or dry acid precipitation) in remote areas far downwind from their point of origin. In essence, the gaseous pollutants are now being exported from urban and industrial centres to the surrounding countryside and to distant areas downwind.

This is the 'tall stacks paradox' (Likens and Bormann 1974) – the inadvertent replacement of an original visible, local air pollution problem (soot) with an invisible long-distance (even trans-frontier) problem (acid rain). The symptoms of the cure may turn out to be worse than the symptoms of the original malaise. Similar explanations of the links between air pollution control initiatives and acid rain are offered in the United States, where 'well intentioned regulations may unwittingly have aggravated the problem' (La Bastille 1981: 662). New rules introduced by the Environmental Protection Agency (EPA) in the 1970 US Clean Air Act caused industrial plants and power stations to increase the height of stacks, so that by 1981 there were 179 stacks at least 150 metres high, of which 20 were as high as 300 metres.

ALARM BELLS RING IN BRITAIN

What had started out as essentially local air pollution problems were being turned into regional problems by the tall stacks policies. By the early 1970s, there were signs of declining interest in local smoke problems and growing interest in the broader distributions of pollution, particularly from sulphur dioxide.

In late July and August 1972, for example, 15,000 school children throughout Britain helped with an air pollution survey organized by the Advisory Centre for Education (ACE) and the *Sunday Times* (*The Times*, 19 June 1972). They collected information on the distribution of selected species of lichen (which are sensitive to sulphur dioxide – see chapter 5). The lichens were used as a 'doomwatch dial' of air pollution, and the aim was to build up a map of levels of pollution across the country. The ACE survey produced alarming results (Mabey 1974: 4). The map that emerged showed only four areas in Britain with pure air – Hadrian's Wall, the Lake District fells, Exmoor, and parts of Norfolk. Particular 'black spots', or lichen deserts, were found in the Thames Valley around London, around Birmingham, in the coalfield areas of West Yorkshire, Durham, and Northumberland, in parts of Scotland's Midland Valley, and in South Wales. These results were all the more alarming because the badly polluted areas were almost identical to those named by the Beaver Committee nearly two decades (and two Clean Air Acts) previously.

Surveys by school children can be ignored by those who question the rigour and scientific objectivity of their results. However, when the scientific 'professionals' enter the scene, results are more difficult to overlook. The regional extent of SO_2 pollution in Britain was unquestionably demonstrated, to even the most ardent sceptic, in studies of rainfall acidity in Scotland by staff of the Institute of Terrestrial Ecology (ITE) published in mid-1982 (Fowler *et al*. 1982). The ITE study showed that rainfall acidity in parts of Scotland was up to ten times higher than normal background levels, and that acid rain was just as much of a problem there as it was in parts of Scandinavia and North America – areas with an established history of this form of pollution. The worst-hit areas appeared to be along the west coast, north of Glasgow, up to the Isle of Skye. Although the survey found that industries in Glasgow appeared to produce up to one-third of the acid rain that fell locally, across Scotland as a whole the pollutants appeared to come from throughout the United Kingdom and even from as far away as mainland Europe.

Within three months of the appearance of the ITE results for Scotland, evidence emerged that acid rain was falling in the rest of Britain. A report commissioned by the Department of the Environment (DoE) and prepared by the UK Review Group on Acid Rain, based at the government's Warren Spring Laboratory, concluded that in some parts of the UK the severity of the acid rain problem was 'of the same order as in other high input areas of Scandinavia and North America' (1982).

BRITAIN'S INVISIBLE EXPORT

The next link in the chain was the rising tide of opinion that argued that not only was Britain fouling its own nest with acid rain, but it was exporting invisible oxides of sulphur and nitrogen abroad. In 1971 the Swedish Ministry of Foreign Affairs first accused its European neighbours of unwittingly dumping their noxious oxides downwind in Scandinavia (Royal Ministry of Foreign Affairs 1971), and the accusations were repeated in 1972 when the Swedish government proposed action to avert acid rain damage to the United Nations Stockholm Conference. Britain was singled out at the time as perhaps the largest single contributor to Sweden's acid rain problem, and the finger was pointed at the Central Electricity Generating Board (CEGB) as the primary culprit. Despite protestations from the CEGB (1979) that it was being wrongly accused – a storyline that was to continue up to the present day (see chapter 10) – the serious allegations were not withdrawn.

By June 1976 Jim Callaghan's Labour government had to confess that sulphur dioxide discharged from power station chimneys in Britain was being deposited as acid rain over Norway and other parts of southern Scandinavia. This startling admission, which has never since been repeated in public – especially since Margaret Thatcher's Conservative government took office in May 1979 – appears in a report by the DoE's Central Unit on Environmental Pollution (Department of Environment 1976). The report recognized that the CEGB's tall stacks policy had dispersed the pollution nuisance to the upper atmosphere but also allowed it to be carried to other countries. This was to be the first of a long line of British reports (see chapter 10) to conclude that the precise effects of acid rain on forest and freshwater resources are not clear, and that many topics need to be further investigated.

The DoE report was a timely forerunner of a more detailed scientific study of sulphur dioxide pollution carried out by the Organization for Economic Co-operation and Development (OECD), which claimed to give rough indications of the extent of the SO_2 problem and was published in Paris in July 1977. Eleven countries participated in the five-year study (1972–6) – Austria, Belgium, Denmark, Finland, France, West Germany, the Netherlands, Norway, Sweden, Switzerland, and the United Kingdom. The OECD survey established that in the early 1970s the eleven countries between them were injecting around 9 million tonnes of sulphate into the atmosphere each year, although the contributions varied considerably from country to country (Table 1.1). All countries received back on their own territories a fair proportion of what they sent up, but some (especially Denmark and the United Kingdom) were net exporters whilst others (especially Austria, Finland, Norway, Sweden, and Switzerland) were net importers. The report concluded that there was strong evidence that the sulphur dioxide pollution caused the acid rain and snow that appeared to be killing salmon and trout, especially in Norway and Sweden but possibly also in Scotland.

Britain was heavily implicated in the OECD report, being described as 'most responsible for polluting Europe's air with sulphuric fumes' (*The Times*, 13 July 1977, p. 7). The study found that Britain not only produced more sulphur pollution than any other European nation, but more of it ended up in the skies above other countries. Britain was producing around

Table 1.1   *Estimated sulphur dioxide emissions in Europe, 1973*

Country	Population (million)	Annual SO_2 emission	
		Per inhabitant (kg)	National total (tonnes)
Austria	7.4	29.6	221,000
Belgium	9.6	51.7	499,000
Denmark	5.0	62.2	312,000
West Germany	61.2	32.1	1,964,000
Finland	4.7	58.3	274,000
France	49.5	32.7	1,616,000
Netherlands	13.2	29.5	391,000
Norway	3.9	23.5	91,000
Sweden	7.9	52.0	415,000
Switzerland	6.2	12.1	76,000
United Kingdom	54.2	51.7	2,803,000

Source: Organization for Economic Co-operation and Development (1977).

2.8 million tonnes of sulphuric pollutants per year, whilst Norway – the country the worst affected by fallout – produced only 91,000 tonnes itself and received a further 60,000 tonnes from Britain.

What at the time was referred to as 'sulphur rain' was being seen as 'one of Britain's least welcome invisible exports' (*The Times*, 28 November 1977, p. 2). Behind the scenes, government scientists in Britain were starting to look carefully at the whole issue and politicians were beginning to realize the implications (both domestic and international) of the recent scientific findings.

By the end of 1977 biologists and chemists from the CEGB were engaged in studies of sulphur damage in Norway, in collaboration with scientists of Norway's Fish and Forests Project (correspondence from Dr Peter Chester to *The Times*, 7 December 1977, p. 19), although British critics suspected that the CEGB was 'up to the classical ploy of using a flurry of research into effects as a smokescreen behind which causes continue unabated' (correspondence from Professor D. Bryce-Smith to *The Times*, 15 December 1977, p. 26). The same accusation was to be repeatedly levelled against both the CEGB and the British government over at least the next eight years (see chapter 10). Public interest in the 'sulphur rain' problem was also rising, and in late November representatives of 20,000 of Britain's youth visited the DoE and the CEGB to ask them to reduce sulphur emissions from British power stations and industry (*The Times*, 28 November 1977, p. 2).

By late 1977 Scandinavian concern over Britain's apparent air of calm over acid rain was mounting. Sven Larsson of Faltbiologerna (the Swedish Youth Federation of Field Biologists) was able to give an alarming inventory of the problem in Sweden (*The Times*, 28 November 1977, p. 2) – 10 per cent of the country's 100,000 lakes were already damaged by acidification, fish could no longer survive in 2,300 lakes in Norway and Sweden, Sweden claimed to have lost around £5 million a year in forest production (and this could rise to £40 million by the year 2000), and the cost of corrosion in the country could be around £250 million by 1981.

Britain was starting to appear as the sleeping giant in the tale, and the goodwill won in past successes in controlling air pollution was rapidly being eroded by the government's attitude of complacency.

AN INTERNATIONAL PROBLEM

It is true that there are immense difficulties in unequivocally establishing the links between cause and effect (e.g. that SO_2 emissions do produce acid rain, and that acid rain does have the inferred effects on lakes, forests, crops, and buildings). Such difficulties are explored in chapter 3.

However, geographical patterns are part of 'the acid test', and these point quite clearly in a given direction. Gross patterns provide a key. Since the mid-1960s, the area affected by acid rain in the Northeast United States, in northern Europe, and in Scandinavia has more than tripled, and these are the regions most exposed to emissions of sulphur and nitrogen oxides (Barabas 1983). More localized patterns reinforce the picture. Downwind, thousands of trees are dying, and lakes and streams that once contained many fish now contain few; hundreds of kilometres upwind, factories and power stations that burn coal or oil are emitting sulphurous gases high into the atmosphere, precisely because they can then be carried long distances before they fall to the ground again (in dust, rain, and snow) (*The Times*, 10 January 1984, p. 2). Granted the connection is mainly one of inference, but this is often all that is revealed in complex environmental problems.

An additional dimension to the acid rain problem in part explains Britain's reluctance to act on the evidence available. It is clear that acid rain has a truly trans-national character, so that the cost of removing a source of the pollutants (mainly SO_2 and NO_x) might well fall in one country while the cost of not removing it – or, conversely, the benefits of removing it – may fall in one or more other countries. Consequently those countries that import acid rain (Austria, Finland, Norway, Sweden, and Switzerland in Europe, and Canada) have a stronger vested interest, and an attitude of much greater urgency, in ensuring that remedial actions are taken in those countries that export it (Denmark and the United Kingdom in Europe, and the USA in North America). It is not necessarily in the best economic interest of the exporters to take precipitate steps (such as the imposition of regulations or emission standards) the economic and broader benefits from which would pass mainly to the importers. This polarization of attitudes is well illustrated in the reply of Mr Patrick Jenkin (the then Secretary of State for the Environment) to a charge of complacency (in Britain's attitude to a draft EEC directive on SO_2) made in the House of Commons on 8 February 1984. He argued that 'the German Government, which clearly feels a greater need because of the problem of the effect on trees to move faster, is going ahead; but we shall need to discuss all this in the context of the Community's directive' (*The Times*, 9 February 1984, p. 4).

The events of Monday, 18 March 1985, on both sides of the Atlantic, highlight the urgency and seriousness of international concern over acid rain. In London, leaders of environmental groups from several European countries including West Germany, the Netherlands, Sweden, and Austria met Mr William Waldegrave (Under-Secretary of State at the Department of the Environment) to protest at the British government's failure to curb acid rain (*The Times*, 19 March 1985, p. 4). Feelings were running high that Britain had refused (and continues to refuse) to ratify a proposed EEC directive calling for a reduction of 60 per cent in sulphur dioxide emissions and 40 per cent in emissions of nitrogen oxides by 1995, and declined to join the so-called '30 per cent club' – a group of industrialized nations committed to reduce SO_2 emissions from power stations and factory chimneys by 30 per cent within the next decade. Some 3,000 miles away in Quebec City, Canada, the Canadian prime minister (Mr Brian Mulroney) was beginning a summit meeting with America's President Reagan. Acid rain – alleged to drift north-west from America's main industrial belt and fall over Canada – was top of the summit agenda, ahead of research into 'Star Wars' technology and the testing of cruise missiles over Canada (*The Times*, 19 March 1985, p. 5).

The same messages emerged from both meetings – the government was reluctant to spend vast sums of money on large-scale clean ups, or to jeopardize anticipated economic recovery by cracking down on industry, without much stronger evidence to link acid rain to emissions of SO_2 and NO_x, to link proposed reductions in emissions with expected improvements in air (and rain) quality, and to link the massive costs of action with measurable benefits (such as improvements in human health and environmental quality, and/or increases in agricultural and forestry productivity). Both sets of critics were told that more research on the subject was required before remedial action was justified either as necessary or likely to be effective (or, more precisely, cost-effective).

CONCLUSION

The burning of fossil fuels in power stations, industry, and motor vehicles appears to be turning Shakespeare's 'majestical roof' into a 'foul and pestilent congregation of vapours', more commonly referred to as acid rain. Having set the acid rain into context, we turn next (in chapter 2) to a more detailed treatment of the processes believed to be involved in creating, dispersing, and depositing the acid pollutants.

2 SOURCES, PATTERNS, AND PROCESSES

'The answer is blowing in the wind'

O clouds, unfold!

(William Blake, *Milton*, preface)

The acid rain debate is complex and founded on many unresolved scientific puzzles. There are two key areas of uncertainty. One is the link between possible sources of sulphur and nitrogen oxides (especially man-made ones) and the production of acid rain. The other is the link between acid rain and damage to the environment. Unless these links are established to an extent and in a way that is convincing – particularly to politicians, the electricity generating utilities, and government scientists – it will be extremely difficult to persuade them that expensive measures to limit emissions from power stations, factories, and motor vehicles will be either cost-effective or necessary.

In this chapter we shall examine how acid rain is formed, and reserve until Part II consideration of the question 'does acid rain really cause the environmental problems widely attributed to it (and if so, how)?'

This lack of complete understanding of how acid rain forms fuels the argument over how to reduce acidity and provides a convenient defence for those (like the British and American governments) who are reluctant to comply with the Geneva Convention. The two camps within the rather polarized debate adopt radically different stances on this critical area of uncertainty (La Bastille 1981: 655). One group argues that we must understand the processes involved before taking action, and therefore the most expedient solution at present is to carry out detailed scientific studies to fill in the gaps in our understanding of the problem. This tactic has been adopted recurrently in Britain (see chapter 10). The other group argues that it will take years to produce unequivocal evidence, during which fish will continue to die, trees wilt, and lakes acidify. They argue that, because there is adequate (if not perfect) evidence to link sources with patterns and effects, emissions (especially from power stations) should be reduced without further delay.

The debate centres around acidity in rainfall, and the effects of this acidity after it lands on vegetation, soils, lakes, and rivers. The concern is mainly over acidification of the environment, for which acid rain acts as the primary catalyst. Before we consider how acid rain is formed, it will be useful to clarify what acidity is and how it is measured.

ACIDITY – DEFINITION AND MEASUREMENT

All elements are composed of atoms, which are made up of stable elementary particles called electrons. Electrons orbit the nucleus of the atom in numbers equal to the atomic number of the element. Hydrogen has the atomic number 1, so it has one electron; oxygen, with the atomic number 8, has 8 electrons; and so on. The structure of an atom is normally stable, but if groups of atoms of different elements come in contact (normally by mixing of the elements) chemical interactions can occur in which one or more electrons are transferred from one atom to the other. In this way ions (electrically charged atoms or groups of atoms) are formed by the loss or gain of one or more electrons. Those that lose are positively charged and called cations; those that gain are negatively charged and called anions. Hydrogen ions are formed from hydrogen atoms that have lost their electrons and become positively charged (i.e. cations).

A dictionary definition of acid is 'any substance that dissociates in water to yield a sour corrosive solution containing hydrogen ions' (Collins 1982: 12). Acids in solution turn litmus paper red. The opposite of an acid is a base, a solution of which is an alkali (basic or alkaline as adjectives). Alkaline solutions turn litmus paper blue.

Although we tend to think of acids as solutions (such as nitric acid), an acid is not necessarily a liquid in solution, although it can be dissolved to produce one. Particular acids are in reality cocktails of ions, which can be dissolved. For example, dissolved sulphuric acid and nitric acid consist of hydrogen ions plus sulphate and nitrate ions respectively.

The pH scale

The term 'acidity' is often used to mean the state of being an acid, but it is more properly used to express the amount of acid present in a solution (i.e. the number of hydrogen ions in circulation). This is normally stated as a pH value, where pH is a short-hand version of 'potential hydrogen'.

The pH scale (Figure 2.1) runs from 0 to 14.0, where 7.0 is neutral. Pure water has a pH of 7.0, acid solutions have a pH less than 7.0 and alkaline solutions a pH greater than 7.0. The higher the pH value, the greater the alkalinity; the lower the value, the stronger the acidity. The pH of some common household substances is shown in Figure 2.1, for comparative purposes. Vinegar, with a pH of around 3.0 is a fairly strong acid; soap solution, with a pH of around 10.0, is a fairly strong alkali.

A growing number of scientists favour the use of hydrogen (H^+) ion concentration (a linear scale) to the logarithmic pH scale: 'the use of a linear representation of concentrations provides the most appropriate means of representing the magnitude of any change in concentration . . .

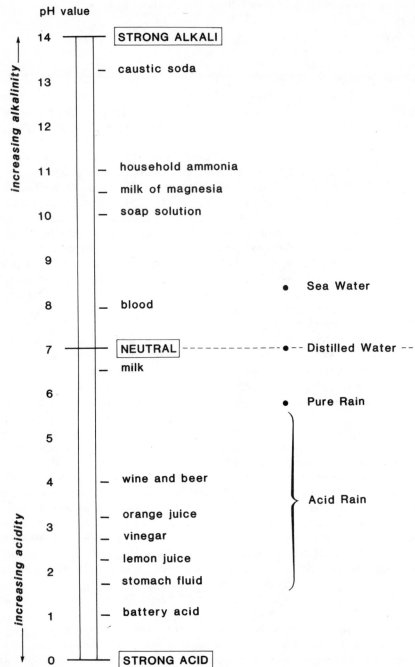

Figure 2.1 The pH scale, with the pH of some common substances

[and] . . . a widespread adoption of the linear concentration scale when reporting H^+ ion concentrations, especially in acidic rainfall and fresh-waters, would appear to be overdue' (Laxen 1984: 409). The pH scale is a useful standard for measuring levels of acidity, but apparently small differences in quoted pH values can be very deceptive because the scale is logarithmic. The pH value is the logarithm of the reciprocal of the concentration of hydrogen ions. What this means in practice is that relatively small differences in pH represent potentially large differences in acidity. Thus, for example, a solution of pH 6.0 is acidic, but a solution of pH 5.0 is ten times as acidic and a solution of pH 4.0 is a hundred times as acidic.

RAINFALL CHEMISTRY

Although pure (distilled) water has a pH of 7.0, water moving through the environment is in contact with chemical elements. Some are produced by natural processes, both biological (e.g. the decay of dead plants) and geological (e.g. volcanic eruptions, and the weathering of rocks). Other elements are produced, normally as unplanned side-effects, by human activities (e.g. air pollution). Through this contact the water is contaminated. Sometimes the water becomes more acidic (e.g. streams draining acid peat bogs) and sometimes it becomes more alkaline (e.g. streams flowing over chalk).

Because of this contamination, normal rainfall is never 'pure' (i.e. pH 7.0). The extent of the impurity varies from place to place, reflecting variations in the source areas of contaminants. Some contaminants (e.g. ammonia, calcium, magnesium, and potassium – Summers 1983: 82) make the rainwater more alkaline, whereas others (e.g. sulphur dioxide and nitrogen oxides) make it more acidic. Most of this variability in rainwater chemistry stems from the distribution of man-made factors such as large cities, industrial complexes, and power stations. This is a key ingredient in the acid rain debate.

It is critically important to establish the background pH of natural (uncontaminated) rain, because this provides the baseline against which increased acidity (caused by air pollution) can be measured and the extent and significance of the increase determined. Although there are spatial variations in the amounts and composition of chemical impurities derived from natural processes (e.g. volcanic eruptions), these are believed to be minor, certainly in comparison with man-made factors.

How clean is clean?

The pH of 'clean' or pristine rain falling in areas remote from industry and other man-made sources of contaminants is generally taken to be 5.6 (Likens and Bormann 1974; Likens 1976; Summers 1983: 87), which is acidic in itself (see Figure 2.1). In fact this is twenty-five times the acidity of distilled water. This normal rainfall acidity, entirely natural in origin, includes weak solutions of carbonic acid (produced from carbon dioxide naturally present in the atmosphere, derived from plant respiration and other sources) and sulphuric acid (produced, for example, from volcanic emissions of sulphur dioxide) (Frohlinger and Kane 1975; Likens *et al.* 1979). This background acidity of rainfall provides nature's free supply of fertilizers for plants and trees, and is an essential mechanism in maintaining the finely tuned chemical balance in environmental systems.

The determination of the normal pH of rain is a vexed question but central to the acid rain debate. Early European rainwater chemistry data suggested that the pH of non-polluted precipitation is above 7.0 (Oden 1976); most scientists take neutral precipitation to have a pH of either 5.7 (Barrett and Brodin 1955) or 5.5 (Mrose 1966).

The Central Electricity Generating Board (CEGB) has a strong vested interest in the matter, having been widely accused of contributing to the acid rain problem in Britain (see chapter 10) and Scandinavia (see chapter 8). It prefers to quote a pH value of 5.0 for pristine rain (1984: 2). This might appear to be a classical tactic of moving the goal-posts while the game is in progress, because the baseline it seeks to be judged against is a hundred times as acid as distilled water, and four times as acid as the generally accepted norm. Some scientists favour the use of pH 5.0 for natural rain, which 'would seem to be more realistic' (Summers 1983: 88) even if it does err on the cautious side.

The CEGB view is *not* entirely without foundation, which comes partly from research carried out in Sweden – paradoxically the country which has been most vocal over acid rain problems in the last decade – at the International Meteorological Institute in Stockholm (Charlson and Rodhe 1982). The Swedish scientists have demonstrated that the natural sulphur cycle (which includes SO_2 emissions from volcanic eruptions and dimethyl-sulphide produced by biological activity on the surface layers of oceans) can produce large variations in space and time in the amounts of sulphur compounds that determine the acidity of rainwater. They found that average pH values in unpolluted regions can be around 5.0 and even fall as low as 4.5 (over twelve times the acidity of pristine rain), entirely through natural processes.

We shall note this significant lack of agreement but adopt the normal scientifically accepted standard pH for clean rain (5.6). In the discussion that follows, 'normal acidity' is taken to have a pH of 5.6, and observed levels are quoted in terms of relative increases in acidity over this background level.

ACIDIFICATION OF RAINFALL

Recent measurements indicate that rains and snows with a pH of 4.3 or lower fall regularly over many of the highly industrial areas of the northern hemisphere, especially North America and northern and western Europe

Table 2.1   *Some reported rainfall acidity values*

Location	Past pH		Present pH	
South-east England	4.5–5.0	(1956)	4.1–4.4	(1978)
Eastern Scotland			4.2–4.4	(1978–80)
West Wales			4.9	(1981–83)
Netherlands	4.5–5.0	(1956)		
West Germany			3.97	(1979–80)
Black Forest (West Germany)			4.25	(1972)
Southern Norway	5.0–5.5	(1956)	4.7	(1977)
Northern Norway	5.5–6.0	(1959)		
Southern Sweden	5.5–6.0	(1956)	4.3	(1975)
Northern Sweden	5.5–6.0	(1959)	4.3	(1972)
Northern Italy			4.3–5.5	(1981)
Switzerland			4.3–5.5	(1981)
Japan (around Tokyo)			<4.5	(1983)
Canada (Quebec)			4.5	(1982)

Sources: Environmental Resources Ltd (1984), Okita (1983), Bobee and Lachance (1983)

(Tolba 1983: 167) (see Table 2.1). The pH of precipitation in southern Scandinavia is generally in the region 4.2–4.3 (Abrahamsen, Stuanes, and Tveite 1983: 89). Such rain is about twenty times as acid as clean rain (pH 5.6), or five times as acid as the normal levels claimed by the CEGB (pH 5.0), and it is held to be responsible for recent acidification of lakes and rivers in many parts of Scandinavia and elsewhere (see chapter 4). The increase is widely believed to stem from man-made emissions of sulphur dioxide (SO_2) and nitrogen oxides (NO_x) released through the burning of fossil fuels which, as we shall see later, can produce strong sulphuric and nitric acids in the atmosphere.

This rain with a pH below around 4.5 is the 'acid rain' on which the debate centres, and from which the acrimony stems. But there is no one precise point along the pH spectrum at which it can be said that 'acid rain' is falling. This is partly because of uncertainty over the pH of clean rain (i.e. where does the problem begin?), and partly because the scale is a continuum with no discrete breaks in it (i.e. the problem is incremental). This gives added complexity to an already complex debate.

There are wide variations in the acidity of acid rain measured at different places and at different times. A number of factors cause these variations, including patterns and rates of oxide emissions, wind speed and direction, chemical interaction in the atmosphere during transport, and processes of precipitation (wet or dry). Observations are now available on the pH of rain in some remote parts of the world (see Figure 1.1), which indicate spatial variations even in the absence of pronounced air pollution. But the evidence also suggests that rain and snow in many industrial regions of the world are now between five and thirty times as acidic as would be expected in an unpolluted atmosphere (Likens *et al.* 1979; Jickells *et al.* 1982). There are three large regions with an annual rainfall pH of below 5.0, and these lie over or downwind from major source areas of sulphur dioxide (SO_2) and nitrogen oxides (NO_x) in north-west Europe, eastern North America and Japan (these areas account for about half of the total man-made emission of SO_2 around the world; Summers 1983: 87).

North America

One of the main core areas of acid rain is Eastern North America, hence the keen interest in the issue in the United States and Canada (chapter 9). In 1974 it was reported that the annual average acidity of rain and snow falling over the Northeast United States was around pH 4.0 (forty times normal acidity) (Likens and Bormann 1974; Cogbill and Likens 1974; Likens and Butler 1981), and monitoring since the late 1970s in the Eastern United States and much of south-east Canada has uncovered frequent rain in the pH range 4.1 (thirty times the norm) to 4.6 (ten times) (La Bastille 1981: 669).

In many parts of North America summer rains of pH 3.5 (125 times the norm) and winter rains and snows of pH 4.2 (twenty-five times) are *not* uncommon; and the annual average pH of precipitation in southern central Ontario (Canada) varies between 3.95 (forty-five times) and 4.38 (over sixteen times) (Douglas 1983: 76). Results of long-term monitoring of the acidity in precipitation over the Eastern United States (Likens and Butler 1981) suggest that levels of acidity are increasing (i.e. pH is falling) and that the area over which acid rain is falling, centred over the main industrial belts, is expanding (Figure 2.2).

Great Britain

Until the mid-1970s it was believed that acid rains were confined mainly to North America and Scandinavia, but in 1977 the OECD study of long-range transport of air pollutants (Organization for Economic Co-operation and Development 1977) found that the average pH of precipitation was well below 4.5 in much of central and northern Europe, including much of eastern England and Scotland (Figure 2.3).

British official concern over acid rain began in 1982 when it was discovered that rainfall acidity in parts of Scotland was comparable to that in parts of Scandinavia and North America (Harriman and Morrison 1981; Fowler *et al.* 1982). The mean pH of rain between 1978 and 1980 was found to be lowest, at between 4.0 (forty times the norm) and 4.3 (twenty times), in the south-east; pH values in south-west, central and north-east Scotland were found to fall in the range 4.4 (sixteen times) to 4.6 (ten times); the north-west was found to receive the least acid rain, with mean pH in excess of 4.8 (six times) (Figure 2.4). This general pattern reflects the movement of air masses from centres of high SO_2 emissions (mainly the major industrial regions of southern Britain and continental Europe) to the south

Figure 2.2 Increase in acidity of precipitation over the Eastern United States, 1955–6 to 1972–3

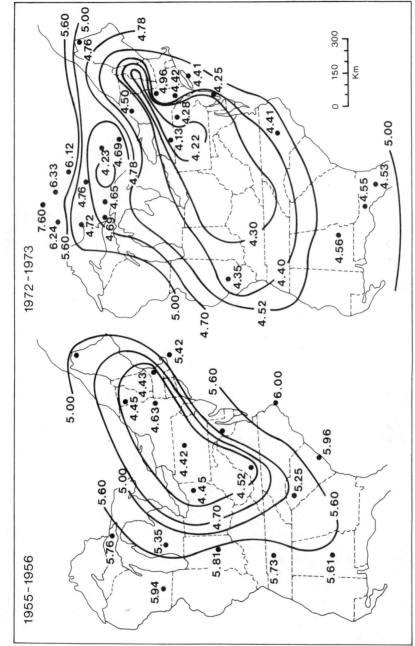

Source: after Likens *et al.* (1976)

Figure 2.3   Variations in acidity of precipitation over Europe, 1974

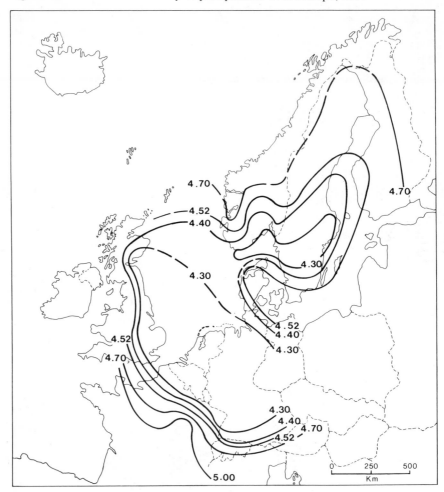

Source: after Organization for Economic Co-operation and Development (1977)

and east of Scotland. British concerns were raised further when the Warren Spring Laboratory – the government air pollution watchdog – found evidence of acid rain (pH below 5.6) falling over parts of northern England, the Midlands, the South, and parts of Wales (UK Review Group on Acid Rain 1982). It is perhaps surprising that more attention was not paid to the results of Gorham's earlier (mid-1950s) studies of rainwater chemistry in the Lake District (mentioned in chapter 1), because about half of his samples – of supposedly unpolluted air – were more acid than pH 4.5 (Gorham 1955).

However, considerable variability is apparent in rainfall pH trends through time, even over relatively short distances. For example, observa-

Figure 2.4 Distribution of mean pH of rain in Scotland, 1978–80

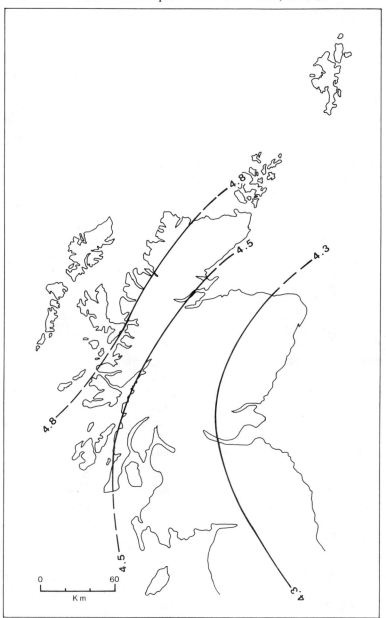

Source: after Harriman and Morrison (1981)

tions at Hawkshead in the English Lake District show that rainfall acidity has not increased over the last twenty-eight years, whereas similar observations just over the Scottish border at Eskdalemuir show an increase in rainfall acidity over the same period (Taylor 1985).

Individual storms (with optimum favourable conditions) have shown some alarming results. American records include a rain of pH 2.7 (nearly 800 times normal acidity) at Kane in Pennsylvania, and a rain of pH 1.5 (over 12,500 times the norm) falling over Wheeling in West Virginia in 1979 (La Bastille 1981: 669). The pH of rain at Banchory in north-east Scotland is sometimes as low as 3.5 (125 times the norm), although the British record presently rests with Pitlochry, where the pH of rain falling in one storm during 1974 was 2.4 (over 1,500 the norm) (Last and Nicholson 1982).

The areas more widely affected by acid rain tend to be located downwind of major cities and industrial areas. As outlined in chapter 1, spatial and temporal patterns of association between observed patterns of rainfall acidity and possible source areas of sulphur and nitrogen oxides (especially large power stations and cities) strengthen the case forwarded by those who are convinced that the man-made emissions of SO_2 and NO_x ultimately cause acid rain.

THE ACID RAIN COCKTAIL

The two basic ingredients of the acid rain 'cocktail' are clear and unchallenged. One is that sulphur dioxide (SO_2) and nitrogen oxides (NO_x) in the atmosphere dissolve in rainwater, and the other is that they are oxidized to produce acid rain or fall as dry deposition (as gases or on dust and other particles). In fact these basic facts have been understood for over a century: R. A. Smith (1872) observed that damage done to plants by acid gases in Victorian Manchester was more spatially restricted during periods of rain and believed that this was due to more effective removal of pollutants by droplets of rain.

Much less clear, and more widely challenged, are the answers to three key questions:

- How much of the oxides is man-made (and where do they come from)?
- How are the man-made oxides released, dispersed, and deposited?
- What role in forming acid rain is played by other factors, such as ozone, hydrocarbons, and other pollutants in the atmosphere (and where do these come from)?

In other words, the debate is *not* over whether SO_2 and NO_x are involved, but over the levels, the most likely source areas, and the best means of reducing the oxides involved.

NATURAL AND MAN-MADE EMISSIONS OF SO_2 and NO_x

It is not really known with any degree of precision exactly what quantities of sulphur and nitrogen oxides exist in the atmosphere around the globe, or

Table 2.2   *Estimated global emissions of sulphur and nitrogen oxides (in millions of tonnes per year)*

Sources	Sulphur dioxide			Nitrogen oxides
	N. Hemisphere	S. Hemisphere	Total	Total
Natural:				
Biogenic	54	44	98	10
– oceans	(22)	(28)	(50)	
– land	(32)	(16)	(48)	
Non-biogenic	22	27	49	10
– volcanoes	(3)	(2)	(5)	
– sea spray	(19)	(25)	(44)	
Total	76	71	147	20
Man-made:				
Fossil fuels	98	6	104	20
Biomass burning	–	–	–	10
Total	98	6	104	30
Total	174	77	251	50

Source: Scriven (1985)

what fraction of the total budget of each is derived from man-made sources. But recent estimates (Table 2.2) offer a rough guide, which is hopefully accurate in relative terms between natural and man-made sources and in absolute terms to within an order of magnitude.

Sulphur

Sulphur is naturally present in the environment, even in the absence of human activity (Deevey 1970; Clapham 1973; Friend 1973; Brown 1982). There are three main natural sources. Most comes from the seas and oceans (in sea salt and gases), and much smaller total amounts come from volcanic eruptions and from natural soil processes (via the decay of organic matter). Although volcanoes contribute relatively little overall, individual eruptions can emit vast quantities over short time-periods in limited areas. For example, sulphur compounds from the Mount St Helens eruption in the USA on 18 May 1980 were converted within a month (by atmospheric processes) into between 0.3 and 0.6 million tonnes of aerosols (Burroughs 1981). This is about twice the normal atmospheric loading for periods with no volcanic activity.

It is difficult to calculate exactly how much sulphur is derived from natural sources, because they are so variable and widely distributed (oceans being the major source). Estimates vary between 78 and 284 million tonnes of sulphur dioxide per year (Tolba 1983: 116). Similarly, sources of man-made emissions (such as power stations, industrial complexes, towns, and cities) are so widely dispersed that exact levels of emission are difficult to determine. Estimates vary between 75 and 100 million tonnes of SO_2 a year. In global terms, therefore, man-made and natural emissions of sulphur are roughly of the same order of magnitude, although the latter exceed the former by up to a half (Table 2.2). But levels of emission vary through time, especially from man-made sources. For example, SO_2 emissions in Europe doubled between 1950 and 1984, whilst UK emissions fell by 50 per cent over the same period; the UK contribution to total European emissions fell from 26 per cent (1950) to around 11 per cent (1984) (*The Times*, 12 June 1984).

The three principal man-made sources of SO_2 emissions are:

(a) The burning of coal (which produces around 60 per cent);
(b) The burning of petroleum products (around 30 per cent); (a) and (b) is done to heat domestic, industrial, and commercial premises, and to generate electricity. It is estimated that around three-quarters of the sulphur emissions in EEC countries are derived from the burning of fossil fuels in industry and power stations (Tolba 1983: 116), and that the CEGB produces around two-thirds of all the SO_2 emitted in Great Britain (Bryce-Smith 1985).
(c) Various industrial processes (around 10 per cent) (Cullis and Hischler 1980). These include the smelting of iron and other metallic (including zinc and copper) ores, the manufacture of sulphuric acids, and the operation of acid concentrators in the petroleum industry.

The man-made sources of SO_2 are all 'point sources' (such as power station and factory chimneys), so there is pronounced geographical variability in atmospheric concentrations of SO_2, even over small areas, especially close to emission sources.

Nitrogen

Data on emissions of nitrogen oxides are much less complete than those for sulphur (McCarl, Raphael, and Stafford 1975). Estimates of natural emissions of nitrogen oxides vary between 20 and 90 million tonnes per year (Tolba 1983: 116), and again natural and man-made emissions are of comparable magnitude (Table 2.2).

The main natural sources of nitrogen include lightning, volcanic eruptions, and biological processes (especially microbe activity) (Clapham 1973: 40–2). Most of the man-made emissions come from power stations, vehicle exhausts, and industrial chimneys (see Table 2.3). Power stations play a much smaller role in the production of NO_x than they do in producing SO_2; it is estimated that the CEGB produces about one-third of Britain's NO_x (Bryce-Smith 1985). Again, most man-made emissions derive from point sources, so that marked spatial variations in NO_x levels are to be expected. Peak NO_x loadings are observed over cities, and levels vary significantly with character and types of transport system, and amount of traffic (Robson 1977).

Table 2.3 *Estimated relative contributions to UK emissions of various air pollutants, 1980*

Sector	Sulphur dioxide %	Smoke %	Nitrogen oxides %	Carbon monoxide %	Hydrocarbons %
Domestic	5	63	23	6	6
Commercial and public	4	2	3	4	5
Power stations	61	6	46	1	1
Refineries and other industry	28	13	—	—	—
Transport	2	16	28	89	42
Non-combustion processes	—	—	—	—	46

Source: Scriven (1985)

The overall levels of NO_x are small in comparison with SO_2 (Table 2.2), although their significance to the production of acid rain may be disproportionately high.

Spatial dimension

These figures on the production of SO_2 and NO_x represent global net budgets and they mask noticeable regional variations, particularly in man-made emissions, which are concentrated into a few heavily populated and industrialized regions (eastern North America, Europe, and Japan; see Figure 1.1).

There is a marked difference in man-made SO_2 emissions between the generally under-developed southern hemisphere and the more highly developed northern hemisphere (with abundant large cities and industrial centres) (Table 2.2). In the south, natural emissions outweigh man-made ones by about 12:1, whilst in the north man-made emissions exceed natural ones by around one-third. Around 90 per cent of all fossil fuel consumption is concentrated in the northern hemisphere. This is reflected in ice core data, which indicate a two–three-fold increase in the deposition of sulphates and nitrates in Greenland over the last century, with no detectable change in the Antarctic (Wolff 1986).

Within the north, the variations are even more pronounced. In 1980, for example, man-made emissions of SO_2 were around 26 million tonnes in the United States, 5 million in Canada, and 70 million in Europe (La Bastille 1981: 660–1). It is estimated that up to 80 per cent of Britain's atmospheric sulphur comes from man-made sources, and this figure rises to 90 per cent for mainland Europe and North America (Galloway and Whelpdale 1980). To make the budget balance in global terms, therefore, man-made emissions over non-industrial regions *must* be much smaller than natural emissions.

Time dimension

Present-day rates of emission are the culmination of a long period of increase, starting with the industrial revolution and reflecting the increasing burning of fossil fuels (see chapter 1). It has been estimated that total European emissions of SO_2 around 1850 were less than 1 million tonnes per year; this rose to around 10 million tonnes by the 1930s, and rates have risen exponentially in the post-war era (Brimblecombe 1985: 11). Moreover, it is estimated that – in the absence of extensive emission controls – SO_2 emissions in Europe are likely to increase by around one-third between 1982 and 2002, with most of the increase concentrated in the south and south-east (Highton and Chadwick 1982).

However, there are signs that this expected continued rise in SO_2 emissions has not fully materialized in recent years. In the UK, for example, emissions rose sharply between 1900 and about 1970, since when they have declined to a level comparable with that of the 1940s (Scriven 1985: 19) (Figure 2.5). United Kingdom sulphur emissions are believed to have fallen by around one-third between 1970 and 1985, with most of the drop occurring in industrial (60 per cent fall) and domestic (50 per cent fall) emissions and a limited and rather erratic fall in emissions from power stations (see Figure 2.5). UK NO_x emissions appear to have peaked in 1979, with a subsequent fall attributed to economic recession and fuel switching (mainly within industry) (Buckley-Golder 1985).

The emerging belief is that in many – but not all – developed countries (including Britain and the United States) SO_2 emissions are not likely to rise further in the coming decades (Tolba 1983: 116), as a result of three trends: better pollution control, the burning of less fossil fuel as a result of energy conservation, and the discontinued post-war rise in energy demand brought about through economic recession in the west. Although all European nations have been asked to cut SO_2 emissions by 30 per cent (see chapter 8), Britain has been unwilling to comply (chapter 10), pointing out that Britain's emissions are already on the decline whilst European emissions overall have continued to increase.

Two persistent mysteries

One of the major mysteries in the acid rain debate is why the acidity of rain falling over parts of Britain, mainland Europe, and North America is increasing, at the same time as total man-made emissions of SO_2 appear to be on the decline. Superimposition of these two opposing trends does little to quell the controversy over what (or, more precisely, who) causes acid rain, particularly that which falls over Canada and Scandinavia. The evidence – patchy and inconclusive as it is at present – is *not* entirely

Figure 2.5   Changing levels of man-made emissions of sulphur dioxide in Britain,
1900–82

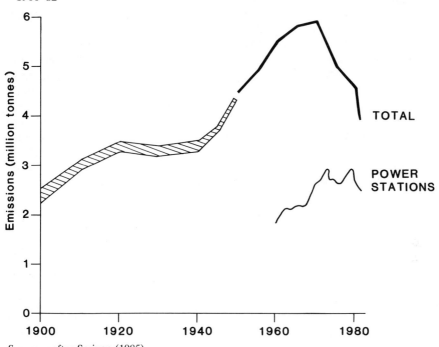

Source: after Scriven (1985)
Note: the hatched area shows range of estimated values.

incompatible with the claim made by the CEGB and other electricity utilities that factors other than just total emissions of SO_2 (such as ozone, hydrocarbons, and atmospheric moisture) must be taken seriously.

A second mystery, of equal significance, is why the acidity of rainfall in some areas shows signs of an increase in recent years, yet other areas downwind of the same possible sources of SO_2 show no such evidence. Part of the answer lies in the unequal availability of reliable measurements of rainfall chemistry across wide areas, and considerable attention is now being directed towards long-term monitoring programmes. Over Europe, for example, a co-operative programme for monitoring long-range transport of air pollutants (the European Monitoring and Evaluation Programme, EMEP) was set up under the 1979 Geneva Convention on Long-range Transboundary Air Pollution (Dovland and Saltbones 1979). In North America, monitoring of acid deposition is carried out by the US National Atmospheric Deposition Program (NADP) and the Canadian Atmospheric Network for Sampling Air Pollutants (CANSAP) (United States–Canada Memorandum 1981). At the global scale, the World Meteorological Office has established the Background Air Pollution Monitoring Network (BAPMON), with over 100 stations collecting data on global trends (Georgii 1981). These programmes are designed to pin-point where and how much acid rain is falling.

Even where records of rainfall acidity exist, the evidence tends to be inconclusive. For example, although there is now little doubt that acid rains are falling over Scandinavia, no significant trend in the sulphate content of Norwegian rain was detected between 1972 and 1981, although the nitrogen content had increased slightly over the same period (Overrein, Seip, and Tollan 1981).

This component of the debate is very complex, and the picture is far from clearly defined. Although both sides in the debate profess to have sound conclusions supported by firm evidence, we must conclude that complexity at present overshadows the prospects of seeing things in black and white. Thus the answer to the first key question appears to be the qualified conclusion that at the global scale natural and man-made sources of SO_2 and NO_x are roughly comparable in magnitude, but over industrial areas in Europe and North America the man-made sources far outweigh natural ones. This leads in to the second question, relating to how the man-made oxides are released, dispersed, and deposited.

FOSSIL FUELS AND OXIDE EMISSIONS

It is widely believed that sulphur dioxide is the main ingredient of acid rain. It normally is, at least by quantity (Table 2.2), although recent evidence suggests that the nitrogen oxides component can produce much more serious ecological damage (e.g. to trees – see chapter 5). The principal source of man-made SO_2 is the burning of fossil fuels. Fossil fuels are 'any naturally occurring carbon or hydrocarbon fuel, such as coal, petroleum, peat and natural gas, formed by the decomposition of prehistoric organisms' (Collins 1982: 572). The organisms, which lived literally millions of years ago, have been converted by natural biological and geological processes – the former include decay and the latter include burial and compaction – into the fossil fuels that were first exploited on a large scale during the industrial revolution.

The fossil fuel paradox

Physical transformation takes place during the long history of burial and compaction, but the fossil fuel deposits preserve within them the chemical elements (including hydrogen, carbon, sulphur, and nitrogen) that were stored in the bodies of the original living organisms and essential for their very existence (Central Electricity Generating Board 1984: 3). When the fossil fuels are burned – mainly in power generation (coal and oil in power stations and petroleum in motor vehicle engines) but also for heat (coal in domestic fires) – the chemical legacy of the past is released into the atmosphere as waste products (pollutants) via chimneys and vehicle exhausts.

Thus arises the 'fossil fuel paradox' – the chemical materials ultimately derived from natural biological processes, which have been 'locked up' for geological time-periods, are released instantaneously back into the environment on combustion. The problem is further accentuated because the fossil fuels accumulate over wide areas, whereas the release through combustion occurs from a large number of individual locations. This rapid release from point sources of what is essentially natural material effectively overloads today's natural environmental cycles, which cannot cope with the quantities involved. Pollution occurs as a consequence.

Combustion and emission

One of the principal ingredients of fossil fuels is sulphur. All fuels contain some sulphur, although the sulphur content varies according to fuel type and source area (Swedish Ministry of Agriculture 1982: 7): crude oil often contains between 0.1 per cent and 2–3 per cent of sulphur; coal has a much wider range, starting at 1–3 per cent of sulphur in coal prepared for 'smokeless zones' (Mabey 1974: 101).

As the fossil fuel is burnt, chemical reactions occur that release the products stored within it (Central Electricity Generating Board 1984: 3). Oxygen in the air at the time of combustion combines with the elements to produce oxides, which are emitted from the chimneys and exhausts. One such oxide is water (hydrogen oxide, or H_2O) in the form of steam. Others include oxides of carbon (especially carbon dioxide, CO_2), sulphur (especially sulphur dioxide, SO_2), and nitrogen (various nitrogen oxides, NO_x), which are released as gases. These exhaust gases can be distributed dry through the atmosphere (on winds), or they can be chemically altered during transport (into acids) and dissolve in water naturally present in the atmosphere (as cloud or rain droplets) to form weak acid solutions (acid rain).

The quantities of sulphur dioxide emitted in burning fossil fuels are not inconsiderable. It is estimated, for example, that a single large American coal-fired power station can emit in a single year as much SO_2 as was blown out by the May 1980 eruption of Mount St Helens – around 400,000 tonnes (La Bastille 1981: 660). The principal industrial source of SO_2 is smelting operations, and these too can emit vast quantities from individual point sources. The 380 metre high chimney at the Inco nickel smelter complex in Sudbury, Ontario (Canada) holds two world records (La Bastille 1981: 662) – as the world's tallest stack, and as the free world's largest single source of SO_2 emissions. In 1981 the complex was producing around 2,500 tonnes of SO_2 a day, down from a previous high of around 7,000 tonnes.

British and American attempts to curb air pollution (discussed in chapter 1) have been based in part on increasing the height of chimneys and installing particle precipitators. Whilst these measures clearly have brought the required improvements by reducing particulate pollution, they appear to have magnified the pollution problem from gases in two ways. The tall stacks policies have meant that smoke and gases are now emitted at greater height, so that they are dispersed (and fall) over a wider area. Moreover, before the air pollution Acts, alkaline particles (fly ash) in the smoke plumes – which were the target of the control measures – served to neutralize the SO_2 and NO_x gases also released from the chimneys. But with the use of devices such as electrostatic precipitators (see chapter 7) and other aerosol-removing technologies the potential acids are no longer neutralized by alkaline particles (Moran, Morgan, and Wiersma 1980).

The switch by some consumers from coal to natural gas as a fuel source – partly promoted by anti-pollution legislation and partly by the changing economics of energy use – may also have aggravated the situation (Tolba 1983). Although natural gas creates less sulphur when it is burned, the sulphur that is produced is *not* neutralized as it had been by the high calcium content of the unfiltered coal smoke emissions.

DISPERSION AND TRANSFORMATION

It is during the process of dispersion through the atmosphere, after being released from chimneys and vehicle exhausts, that the SO_2 and NO_x undergo the chemical transformations that produce acid rain. Many of these involve interactions with materials that are naturally present in the atmosphere (such as sunlight, ozone, and moisture). Other pollutants released from power stations, industrial boilers and smelters, and vehicles (such as acidic soots and small particles) can also play a part in the production of acid rain.

As well as making chemical transformation possible, dispersion spreads the oxides geographically. The result is that oxides emitted from point sources (natural or man-made) are deposited over and can affect large areas downwind. There are two mechanisms of deposition (Garland 1978; Whelpdale 1983):

- dry deposition (in the form of gases and particles), and
- wet deposition (in the form of rain, snow, mist, and fog).

Some of the oxides fall to the ground and land on the surface of vegetation, soils, and lakes either as gases or on small particles. This is dry deposition, which occurs mainly close to points of emission, although long-distance dry deposition does occur. The rest of the oxides are transformed (through oxidation) by atmospheric oxygen into sulphuric and nitric acids. Sulphuric acid cannot exist as a gas; it exists only on small particles or in solution (dissolved in cloud and rain droplets, on snow, or in fog). A variable cocktail of both acids will then fall with precipitation in the form of rain, snow, or mist (occult precipitation – see pages 47–8). This is wet deposition (or, quite literally, acid rain), which tends to fall long distances downwind of the points of emission of the original oxides. This complex brew of transformed and dissolved oxides represents a secondary form of pollution, in contrast to the primary pollution of direct dry deposition of (largely unchanged) oxides.

Chemical complexity

The chemistry of dry and wet deposition is extremely complicated, because a large number of interactions are involved. Moreover, the detailed chemical composition of 'acid rain' varies from place to place, depending upon the quantities of the different oxides involved and the chemical transformations that they have undergone during dispersion in the atmosphere. Rain with a pH of 4.5, for example, may contain a high content of sulphur, a high content of nitrogen, or any combination of the two. Moreover, since other oxides exist naturally in the atmosphere and are released by human activities, these too can contribute to rainfall acidity to a greater or lesser extent. Strictly speaking, therefore, 'acid rain' is a descriptive term embracing a wide variety of mixtures of acids and oxides. There is no one 'acid rain', either in strength (pH) or composition (chemistry). However, the sulphate component is normally dominant.

A simple model (Marsh 1978; Mason 1978; Central Electricity Generating Board 1984: 6) of this complex pattern would have six stages (see Figure 2.6):

(1) the atmosphere receives oxides of sulphur (SO_2) and nitrogen (NO_x) from natural and man-made sources on the ground
(2) some of the oxides fall directly back to the ground as gases (SO_2 and NO_x) without delay or long-distance transport, by dry deposition
(3) sunlight stimulates the formation of photo-oxidants (such as ozone O_3) in the atmosphere
(4) these photo-oxidants interact with the SO_2 and NO_x to produce sulphuric acid (H_2SO_4) and nitric acid (HNO_3) by oxidation
(5) the sulphur and nitrogen oxides, photo-oxidants and other gases (including ammonia, NH_3) dissolve in cloud and rain droplets to produce acids (H^+ and NH_4^+), sulphates (SO_4^{2-}) and nitrates (NO_3^-)
(6) acid rain containing ions of sulphate (SO_4^{2-}), nitrate (NO_3^-), ammonium (NH_4^+), and hydrogen (H^+) falls by wet deposition.

Precursors, radicals, and synergism

The most important step in this chain reaction is the catalytic oxidation of sulphur and nitrogen oxides (step 4), a process that may last for a few hours to a few days and requires photo-oxidants as so-called 'precursors' (i.e. the process would not take place without them) (Lesinski 1983: 157). The conversion of SO_2 to sulphuric acid requires two key ingredients – a catalyst and a photo-chemically generated radical (groups of atoms that are found in several compounds). The latter is the photo-oxidant, which is formed under the influence of sunlight, so that sunny weather conditions enhance conversion.

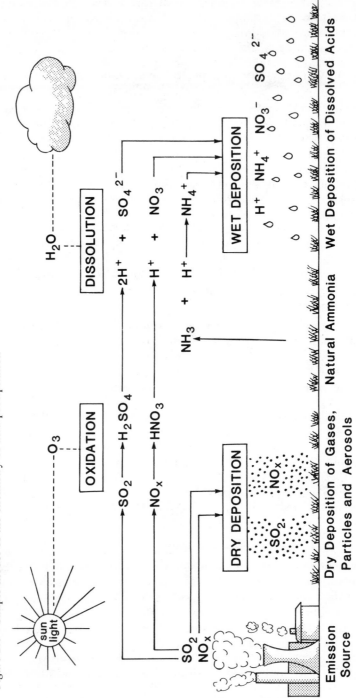

Figure 2.6 Simple model of the chemistry of acid precipitation

The reaction product is a haze of extremely small droplets (mostly less than 1 micrometre in diameter), which are generally invisible. However, the droplets are sometimes visible, when present in high concentrations and under suitable weather conditions, as photo-chemical smog or haze (e.g. arctic haze, pages 57–8).

The chemistry of acid rain illustrates what scientists call 'synergistic effects'. This means that a reaction or set of reactions between two or more ingredients (in this case SO_2, NO_x, sunlight, ammonia, and moisture) produces an effect greater than the sum of the individual effects. Two plus two equals four; but SO_2 plus NO_x plus ammonia plus sunlight plus moisture produce a strange acid brew that is more variable, complex, and potent than all of the ingredients taken separately or simply summed. The final product reflects the photo-chemistry and physical dynamics of the stratosphere (lower atmosphere) as much as it does the initial oxide emissions.

Present scientific understanding of how acid rain forms demands the presence of various catalysts such as ozone, sunlight (to trigger the photo-oxidants), and perhaps also ammonia released naturally from the ground in rural areas (Anderson *et al.* 1985; *The Times*, 12 July 1985). It has even been suggested by some Dutch scientists (Van Breemen *et al.* 1982) that ammonia given off from liquid manure from cows can increase the effects of acid rain (by combining with SO_2 to create fine white crystals of ammonia sulphate, which fall (dry) on plants and trees and are then dissolved and washed off by rainfall into underlying soils).

The ozone connection

The atmosphere is a complex chemistry set, and recent studies (Anderson *et al.* 1985) have shown that around 200 elementary photo-chemical reactions take place within the stratosphere, involving free radicals that may be present in concentrations as low as a few parts per hundred thousand million. Scientists are becoming increasingly aware of the likely significance of the photo-chemical oxidation process in the formation of acids in the atmosphere, and it is now widely accepted that rates of acid generation are critically affected by the presence of oxidizing agents and by the characteristics of the reactions (Calvert *et al.* 1985).

Ozone, one of the most abundant and readily available photo-oxidants, is heavily implicated in the acid rain debate in terms of both observed trends in air quality and detected evidence of damage to trees and plants (see chapter 5). Measurements show a continuous upward trend in ground-level ozone concentrations in Europe in recent decades, which is 'indicative of an increase in chemical reactivity in the atmosphere which (it is believed) will probably be reflected in an increased rate of acid production' (Penkett 1984). Moreover, ozone levels often vary seasonally in response to changing meteorological factors – especially the amount of sunlight to 'drive' the photo-chemical reactions – and this seems to account for the regular seasonal pattern of variations in rainfall pH observed in some areas (Anderson *et al.* 1985).

The two main precursors of ozone are nitrogen and hydrocarbons – both derived mainly from vehicle exhaust emissions.

The linearity debate

The so-called 'linearity debate' can be summarized in the form of a simple question: 'Is the amount of acid emitted into the atmosphere (e.g. over Europe) in any way linked to the amount of acid that falls (over Europe)?' Translated into politics, this means: 'Will money spent on cutting emissions have any practical effect on the ground, or will it all go up in literal and metaphorical smoke?' (Pearce 1985: 10).

In the light of the growing awareness of the possible significance of ozone, there is some foundation to the claim that acid deposition is unlikely to decline in simple proportion to any reduction in the emission of SO_2 and NO_x. The outcome of emission reductions may, as many opponents (such as Britain's Sir John Mason, 1984) argue, be extremely unclear and will doubtless depend heavily on the availability of oxidants such as ozone.

Dispersion, diffusion, dilution

The geography of acid rain is almost as important as its chemistry, because direction of travel (and thus pattern of deposition) is controlled by prevailing wind patterns. The length of time over which the material is transported in wind systems is also of great significance, because the longer the SO_2 and NO_x remain in the air, the greater the likelihood of chemical transformation (oxidation) to produce the sulphuric acid (H_2SO_4) and nitric acid (HNO_3) (Junge 1974; Hegg and Hegg 1978; Summers 1983).

Dry and wet deposition generally occur in different places. Although the former can fall either close to or downwind of emission sources, the latter is only possible some distance downwind (after oxidation and dissolution have taken place). Downwind (i.e. with the passage of time), wet deposition becomes greater than dry deposition. The observed pattern reflects what is happening to the plume of oxides during transport by wind (see Figure 2.7).

The emission plume of SO_2, vented from the high stacks, is dispersed by natural turbulence in the so-called 'mixing layer' (the lower 1–2 km of the atmosphere). Buoyant uplift keeps the plume aloft, and it is carried downwind and often retains a thread-like appearance for 1–2 km. It is normally well mixed with the surrounding air after perhaps 5–25 km, through dispersion triggered off by atmospheric turbulence (Central Electricity Generating Board, 1984: 4).

The direction in which the plume travels, the speed of movement, and the rate of dispersal all depend upon meteorological factors such as wind

Figure 2.7 Simple model of dispersion and deposition of acid substances

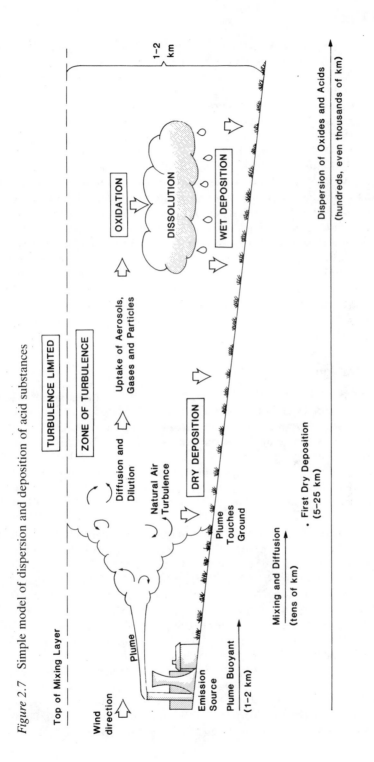

speed and direction, turbulence, air temperature (especially the vertical temperature gradient, i.e. the rate at which the air becomes cooler with increasing height above the ground), and atmospheric stability. Under stable conditions – for example at night over land in all latitudes, and during the day over snow-covered ground (Summers 1983: 83) – there is little if any vertical mixing of air. Under such stable conditions plumes can retain a distinct form for considerable distances downwind.

Except where prevented from doing so by atmospheric conditions (such as inversions, in which cold air lies over warm air and prevents it from rising upwards), the shape of the plume normally evolves by diffusion. It starts at source as a well-defined single thread pointing downwind, and this can often be seen for the first few kilometres. As mixing takes place, the thread changes progressively downwind to a widening cone, which is more diffused and less distinct (Figure 2.7). The rate of diffusion, and hence the rate of dilution of the oxides in surrounding air, is faster when air flow is turbulent (e.g. in flowing around obstructions such as large buildings or mountains). The gases themselves are not visible to the naked eye (though they can be detected using infra-red cameras), and diffusion has by now scattered the smoke and particles over a wide area, so the plume is to all intents and purposes invisible.

Plume tracking

Dispersion in plumes is clearly a key process in the formation of acid rain, and it is understandable that much attention has been devoted to the study of downwind movements of oxides from man-made sources, particularly power stations that emit large quantities of SO_2. The CEGB now has a regular tracking programme using fully instrumented Royal Air Force Hercules aircraft to follow the trajectory of emissions from power station chimneys in Britain en route across the North Sea towards Scandinavia (Cocks, Kallend, and Marsh 1983; Central Electricity Generating Board 1984: 6).

Some critics of the call for urgent action to reduce SO_2 emissions from British power stations point to the lack of clear evidence to link power station emissions with downwind acid rain. Such evidence does none the less exist. Scientists from the United States National Aeronautics and Space Administration (NASA) have succeeded in using airborne laser equipment to follow pollutant materials (SO_2, NO_x and CO_2) from American power stations and industrial plants into the upper atmosphere, where they are converted into acid rain and snow (and into ozone and photo-chemical mists), and from where they are deposited over remote parts of the ocean and over polar regions (*The Times*, 23 April 1984, p. 3). Similar airborne techniques have been used by a team from the Air Resources Laboratory of the US National Oceanic and Atmospheric Administration (NOAA) to study the effects of air pollution on cloud behaviour downwind of the Homer City power station in Pennsylvania.

The NOAA scientists found that polluted clouds had about twice as much moisture, two hundred times as much nitric oxide (NO) and twelve times as much sulphuric acid (H_2SO_4) as unpolluted ones in the same area.

It has also proved possible to track the trajectories of acid-laden air masses from routine general meteorological observations. For example, in February 1984 a distinctive black acid snow fell over remote parts of the Cairngorms in the Highlands of Scotland, and it was possible to trace this to heavily polluted air originating to the south-south-east in industrial regions of central and northern England (Davies *et al.* 1984). The polluted air was believed to have travelled at an altitude of 1.0–1.5 km in association with a stable atmospheric layer.

ACID DEPOSITION

What goes up must come down, and under normal conditions (i.e. away from high mountains and other barriers) the bottom of the plume will generally touch the ground between 5 and 25 km downwind from the emission source (Figure 2.7). Ground-level concentrations of SO_2 gases and particles (dry deposited) are normally at a maximum at this first point of touchdown. This mechanism of producing short-range dry fallout explains why many of the worst symptoms of direct SO_2 poisoning of vegetation (described in chapter 5) – for example on certain lichens – are seen in or close to cities and industrial complexes.

The rest (normally most) of the oxides circulate with the great air masses that form weather systems, to produce what is literally high-flying pollution. The flight of the oxides, during which oxidation and dissolution (in clouds and on raindrops and snowflakes) take place, may last for several days and cover hundreds if not thousands of kilometres. Such long-distance journeys quite naturally show no respect for administrative or national boundaries, and thus the trans-frontier pollution problem arises (see chapter 8).

Particulate material is removed from the atmosphere by the process of scavenging. The speed and efficiency of this process depends on three factors: the height at which particles are injected into the atmosphere, meteorological conditions, and the size, shape, and surface properties of the particles. Emission height (a function of chimney height) is critical, because the higher the particles, the longer they remain in the atmosphere (hence the greater the prospects of chemical transformation) and the more widespread their distribution (hence potential damaging effects).

The popular – and until about 1983 the prevailing scientific – view of acid rain is literally that dissolved acids fall as rain drops. But studies of the air pollution falling on Great Dunn Fell in the Lake District (Dollard, Unsworth, and Harvey 1983) suggest that this view is misleadingly simple and acid mists and fogs must be taken into account. It is now known that high concentrations of sulphates and nitrates exist in the form of fine aerosol particles (dust or soot) in wind-driven ground-level cloud, and that tiny water droplets that condense around these contain much higher concentrations of sulphuric and nitric acids than normal rain in the area. This is termed 'occult precipitation'. Because the droplets are a fraction of the size of normal rain drops, the 'occult' component of acid rain is not measured properly in standard raingauges. Consequently it is under-estimated by up to 20 per cent in published rainfall and acidity figures; 'these underestimates would be most significant in forested regions because wind-driven drops are captured more efficiently by trees than by short vegetation' (Dollard, Unsworth, and Harvey 1983).

ACID SURGES

Sulphur and nitrogen oxides and undissolved acids gradually filter down from the atmosphere and reach the ground surface by dry deposition. Those that fall on lakes and rivers are dissolved, and acidify these areas. Those that fall on vegetation and soils remain temporarily inactive in a chemical form of suspended animation. There they await the first dewfall or rainstorm, which will activate their chemical potency by converting them (by dissolution) into droplets of acid.

During periods of drought or low precipitation, most of the oxides and acids – even long distances downwind of emission sources – fall by this process of dry deposition, accumulate, and subsequently become highly concentrated acids that are quickly washed out by rain and introduce 'acid shocks', 'surges', or 'flushes' into an otherwise stable environment (correspondence from Chris Rose to *The Times*, 20 August 1984, p. 2).

A similar form of suspended animation occurs when oxides and acids are deposited on snow and ice. During winter in arctic areas, pollutants accumulate in and on the snow pack. When spring brings milder weather, the acids concentrated in the snow – which have fallen over the long period of cold and thus are present in large quantities – are suddenly released (over a period of a few days to a week) with the first meltwater. The first meltwater may be five to ten times as acidic as the rest of the snowpack (correspondence from Dr T. D. Davies to *The Times*, 30 January 1985, p. 13). The high acidity of first melt is caused mainly by the sudden release of high concentrations of pollutants on snow melt, but other processes – such as drip from leaves and infiltration into soil (see chapter 3) – may also be important (correspondence from Dr A. G. Thomas to *The Times*, 8 February 1985, p. 10). This produces an acid 'surge' that drastically alters water chemistry in the streams and downstream lakes.

Such sudden and short-lived acid surges, however caused, can have dramatic ecological impacts. Large-scale fish deaths are often associated with the effect of snow-melt acid surges on lakes and rivers (see chapter 4). Surges also affect plants and trees, and at least part of the recent forest dieback in some areas (see chapter 5) may be related to such temporary but disturbing chemical imbalances following prolonged drought (e.g. during 1976 and 1984 in parts of western Europe).

Impact, flux, and variability

The mechanisms by which acid rain is produced are clearly complex. Despite the considerable advances in understanding that recent scientific studies have brought, there are still a great many unresolved aspects of both the science and geography of these processes. There is widespread disagreement, even amongst scientists and specialists, over the choice of most important areas of uncertainty. This spills over into the acid rain debate as played out in public (see Part IV). Concern centres less on the rainfall acidity *per se* than on its impact on the environment. Effects are more important than amounts in the debate, and the ability of an area to assimilate high levels of acidity without serious damage varies considerably from place to place (depending on the buffer capacity of the area, outlined in chapter 3). Effects are the focus of the next four chapters, but two particular aspects of the effectiveness of acid rain are worthy of mention here.

Peak acid loadings

Acid deposition, especially by rain, is highly episodic, and, from an ecological point of view, a few isolated peaks of high acidity can be much more serious than a general background of limited acidity. The extremely low pH values found in some individual rainstorms (some examples of which were noted earlier) indicate that levels of acidity in any one place can be extremely variable through time. This reflects variations through time in such factors as oxide emission rate, wind speed and direction, sunlight, and presence of photo-oxidants in the atmosphere.

Acid surges after long dry spells or during spring snow melt are vivid examples of short but potentially very damaging episodes of high acidity. It has been pointed out, for example, that Norway is no more polluted by sulphur than are many parts of Scandinavia, but snow accumulates there for around six months in the year and acid flushes in early spring cause most of the problems attributed to acid rain (Cadle, Dasch, and Grassnickla 1984; Davies *et al.* 1984; Crawshaw 1986).

Acid flux

A second critical factor is that the effectiveness of acid rain is related to the acid flux (i.e. the total quantity of acid received), not just the pH of rain. The total quantity received at a given place represents the total number of hydrogen ions falling there in both wet and dry deposition. In dry deposition this depends on the concentration of the sulphur and nitrogen oxides that land, which in turn reflects patterns of dispersion and distances transported (Figure 2.7). In wet deposition it depends on rainfall amount and rainfall acidity – the same flux can be produced by small amounts of highly acidic rain or large amounts of weakly acidic rain.

Recent evidence suggests that the total quantities of acid being deposited over the Lake District in Cumbria are high, as much because of high rainfall as through high acidity (low pH) (Taylor 1985; Fowler *et al.* 1986). Similarly parts of Wales now have high acid fluxes, and Dr Alun Gee (Welsh Water Authority fishery scientist) stressed in 1983 that 'it is not so much a problem of acid rain, because the rain here [in Dyfed and Gwynedd] is not particularly acid compared with Sweden and Germany, but we get a lot of rain and it is therefore cumulative in its effects' (quoted in *The Times*, 14 January 1983, p. 4).

Part II

THE SCIENCE
OF ACID RAIN

3 SCIENTIFIC COMPLEXITIES

> Chaos and confusion are not to be introduced into the order of nature because certain things appear to our partial views as being in some disorder.
>
> (James Hutton 1795; quoted by Davies 1968: 162)

The acid rain debate is extremely complex, and it embraces a wide spectrum of themes and issues. The present polarity of views – particularly at the international diplomatic level – reflects inherent complexity, which stems from a unique blend of scientific uncertainty and political misunderstanding of the issues involved. The root causes of the scientific uncertainty form the focus of this Part; the political dimension will be dealt with in Part IV.

There is no simple answer to the apparently simple question 'Does acid rain affect the environment?' The debate is shrouded by many speculations and inferences, often based on scanty or circumstantial evidence. Public interest in the environment has been described as a 'confused mixture of quasi-scientific concern and thorough going sentimentality' (Clayton 1971: 66). This is particularly true of general public interest in acid rain, and it underlines the need to promote a better understanding of the scientific arguments in order to eliminate – in so far as it is possible – the sentimentality.

The aim in this chapter is to explore some of the inherent scientific complexities in the acid rain debate, before we move on in subsequent chapters to examine in detail the possible effects on lakes and rivers (chapter 4), forests and plants (chapter 5), and buildings and humans (chapter 6), which lie at the very heart of the debate.

DAMAGE ATTRIBUTED TO ACID RAIN

Whether we are presently poised on the brink of catastrophic and irreversible environmental damage from acid rain – as some critics argue – is open to debate, but there is little doubt that present-day levels of acid deposition are causing serious damage in some parts of the world. Acid rain is invisible, and its effects may remain undetected even in areas where it has been falling for many years, so that it is extremely difficult to link suspected cause with observed effect.

Whilst there is substance to the argument – staunchly forwarded by opponents to emission control (see Part IV) – that there is little firm evidence to unequivocally link observed damage to the environment with increased acidity of precipitation, the fact remains that 'for over twenty years Sweden and Canada have been looking on helplessly while their lakes and forests die' (Marnot-Houdayer 1985: ix). Moreover, the spatial coincidence between suspected causes and observed effects is striking and this may be a useful diagnostic 'acid test' (Figure 3.1). Most of the damage that has been reported comes from areas that are downwind of SO_2 and NO_x emission sources and receive wet and dry acid deposition. This reflects the dominant trajectories of wind systems bearing oxide-laden air across land and sea with no respect for political or administrative boundaries. There seems little doubt that since the 1950s the use of tall chimney stacks to reduce smoke pollution in and around cities (described in chapter 1) has led to dispersal of SO_2 over much wider areas.

Diffusion of damage

International concern was first aroused at the United Nations Stockholm conference in 1972 by Swedish reports of massive fish kills attributed to acid rain (Swedish Ministry of Agriculture 1982). Since then signs of damage to lakes, forests, vegetation, and buildings, believed to be associated with acid rain, have been reported from many countries. This reflects two super-imposed trends since the mid-1960s: acute damage has doubtless spread over wider areas, and at the same time better surveillance of high-risk areas has uncovered more varied and more subtle forms of damage.

The pattern of reported effects is somewhat patchy, largely because of variations in levels of acid input (via precipitation) and variations in natural buffering capacity. But the 'innocent until proven guilty' argument simply does not apply in the acid rain trial, because the lack of reported evidence from an area does not necessarily mean that acid rain is having no effects there. It may well reflect lack of detailed study; 'no news' might mean quite literally that. As yet only the most sensitive areas and areas where serious ecological problems have arisen – the two are largely the same – have been studied in detail.

Up to now problems have been reported mainly from three areas in the northern hemisphere (Figure 1.1):

- southern Scandinavia (downwind of suspected source areas like the Ruhr and industrial Britain)
- industrial parts of central and western Europe, and
- eastern parts of North America (Canada and the United States, especially in areas influenced by acid precipitation originating in the industrial lower Great Lakes basin) (Likens, Bormann, and Johnson 1972; Likens *et al.* 1979).

Figure 3.1 Spatial coincidence of acid rain, sensitivity, and damage in southern Norway, early 1980s

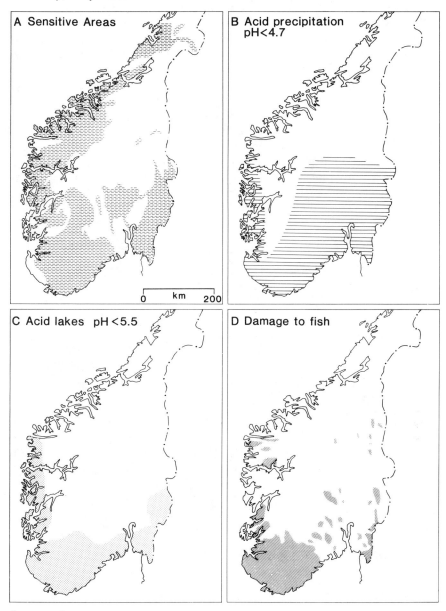

Source: after Wright (1983)

These areas receive large amounts of acid deposition (see Figures 2.2 and 2.3), and their limited abilities to 'absorb' acids (i.e. low buffering capacities) make them particularly sensitive to acidification.

What was once seen as a problem confined mainly to Sweden and Norway has now emerged as a much broader problem affecting nearly every country in Europe (Wright 1983: 137) (Figure 3.2). Most of the evidence is compatible with the argument that oxides emitted over industrial areas and redistributed by major wind currents play a critical – if not the key – role.

Some of the most dramatic effects reported from North America come from Ontario in Canada, where the hard bedrock of the Canadian Shield and poor soil cover offer limited buffering capacity to neutralize the acid rains that fall over the area. The rains are widely believed to blow from America's Midwest industrial states. Canadian fears were well articulated

Figure 3.2   Distribution of damage to lakes and forests in Europe, early 1980s

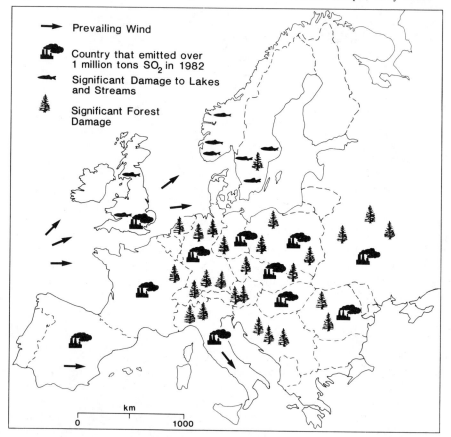

Legend:
- Prevailing Wind
- Country that emitted over 1 million tons SO$_2$ in 1982
- Significant Damage to Lakes and Streams
- Significant Forest Damage

km
0   1000

Source: after Elsworth (1984a)

by the head of the country's Environmental Protection Services who insists that 'we are absolutely convinced it is true, and we are darned worried' (La Bastille 1981: 665). However, the blame might not rest exclusively with the United States because Ontario also has its own sources of oxide emissions, including the Inco smelter complex at Sudbury (Chan *et al.* 1982, 1984; Clark *et al.* 1984) – the free world's largest single emitter of SO_2 (see page 39).

A global problem?

Other industrialized areas that emit vast quantities of SO_2 and NO_x – such as Japan (Okita 1983) – receive large acid inputs (see Figure 1.1) but have yet to show widespread or serious signs of damage to the environment. It is likely to be simply a matter of time before damage is detected in such industrialized areas, unless emissions are checked (both nationally and internationally) before they reach critical levels.

A new dimension to the acid rain problem is emerging. This is the risk of a global distribution of pollution and damage (Tolba 1983: 115), embracing the southern hemisphere and developing countries as well as the industrialized northern hemisphere nations in which most damage has been reported up to the present. Acid pollution has been reported from many developing countries – including Zambia, South Africa, Malaysia, and Venezuela (McCormick 1985) – and in 1982 it was reported that 'there is growing concern in the Third World as sulphur belches from the huge new conurbations of Latin America, India and the Middle and Far East . . . the most productive farmland in China, the paddy fields of south east Asia and the forests of the Amazon may all depend on soils as susceptible to acidification as those in Scandinavia and Germany' (Pearce 1982a).

Some scientists and environmentalists argue that we are presently on the eve of destruction of progressively larger areas of the world's environment, on which we depend for food, water, and natural resources. If that is so, if acid rain is largely responsible and if the acid rain is caused principally by the emissions of SO_2 and NO_x from power stations, factories, and vehicles . . . then the luxury of long-term research studies (see Part IV) can no longer be afforded. Rearranging the deckchairs on the Titanic is *not* a viable solution.

Arctic haze

The significance of the trans-national dimension of acid rain was brought to a head in 1981 with the discovery of extensive acid pollution in the Arctic. Analysis of air samples in Barrow, Alaska – nearly 1,000 km (600 miles) into the Arctic – showed that air pollution is carried (especially during winter and spring) in jet streams in the upper atmosphere, up to 8,000 km (5,000 miles) from the most likely source areas in North America and Europe (*The Times*, 30 June 1982). This evidence of present-day pollution complements ice-core evidence of rising levels of atmospheric sulphates over the Arctic over the last two centuries (see chapter 1).

The Arctic pollution is evident in the form of atmospheric haze, vividly described by Matthew Bean, a 69-year-old Upik Indian from Bethel on Alaska's south-west coast: 'some days you see a bluish-grey haze, which never used to be seen; and some days skies are pale, rather than the dark blue that I used to notice forty years ago' (quoted in *The Times*, 7 October 1984, p. 12). The Arctic hazes are similar in form and origin to the haze layers that frequently occur over and downwind of large industrial areas in Western Europe and Eastern North America. Haze layers often cover a horizontal area of up to 1,000 km (roughly the size of an anticyclone; Bolin and Charlson 1976). They are caused by the scattering of solar radiation by minute suspended particles in the atmosphere, in the size range 0.1–1 micrometre in diameter (mainly sulphate aerosols derived from burning fossil fuels; Charlson *et al.* 1974).

Hazes are a direct product of scattering, and they mainly affect visibility; they are not to be confused with the more complex photo-chemical smogs that occur in some heavily polluted cities (such as London in the past – see chapter 1 – and present-day Los Angeles – see chapter 6) and affect human health. Field studies have shown that the amount of scattering of solar radiation by suspended particles in haze layers is significantly correlated with sulphate concentrations in the overlying air (Waggoner *et al.* 1976), so the link with air pollution is quite clear.

By 1984, as a result of more detailed studies, a better understanding of the Arctic hazes was beginning to emerge. The haze was found to be thickest on Alaska's North Slope, extending eastwards to at least Norway (*The Times*, 7 October 1984, p. 12). This implicated the Soviet Union as a major likely source of the particulate material, and to a large extent pardoned North America. Moreover, general levels of pollution were found to be 'consistently less than in a big city' (K. Rahn, quoted in *The Times*, 7 October 1984, p. 12).

None the less the discovery of pollution in such remote areas, long distances downwind of possible point sources of emissions, highlighted the mobility of the material involved and the ubiquitous nature of its distribution, and reinforced fears of the prospects of global damage from acid deposition.

BUFFER CAPACITY AND ACIDIFICATION

Part of the complexity inherent in the acid rain debate can be attributed to the variability of response of different areas to acid deposition, and the attendant spatial patchiness of evidence of damage. The soils and waters in some areas are able to withstand quite high levels of acid deposition without showing adverse symptoms, whilst other areas (even close by) are not. Thus rain of a given pH may produce detectable effects on the ∢ish, trees, and soils in one area but another area (or another part of the same area) may remain unaffected by the same rains.

The key to unlocking this particular puzzle is the ability of an area to cope with the acid deposition (wet or dry) that falls over it. This is referred to as the 'buffer capacity' of the area. Natural chemical defence systems in the environment offer resistance against acidification. For example, many soils contain minerals (such as lime) that are alkaline, and these will serve to neutralize the acids that percolate through them. The chemical reaction in which negatively charged ions take positively charged hydrogen ions (the definitive property of acids) in hand and effectively neutralize them (Swedish Ministry of Agriculture 1982) is referred to as 'buffering'. Consequently the concentration of buffering substances in a solution is expressed in terms of alkalinity (see Figure 2.1), the inverse of acidity.

There are considerable variations in buffer capacity from place to place. Four main sources of buffering are involved, although they are not all equally significant in all areas.

Natural variations in buffer capacity of soils, deposits, and bedrock

These largely reflect geological factors, particularly the distribution of limestone and other carbonate rocks. Limestone is very alkaline, hence a good buffer; limestone regions are rarely affected by acid deposition. Most other rocks – i.e. carbonate-free softwater areas – are poorly buffered. Areas most vulnerable to acid inputs fall into three main categories (Likens *et al.* 1979):

- glaciated areas on granitic and other highly siliceous bedrock (e.g. granite, certain gneisses, quartzite, and quartz sandstone), with deposits derived from such materials,
- areas with thick deposits of siliceous sands (e.g. the sand plains of Denmark and the Netherlands), and
- areas with relatively old, highly weathered and leached soils.

The regions that have yielded the strongest evidence of damage from acidification are the glaciated Pre-Cambrian shield areas in Scandinavia (especially southern Sweden and Norway), glaciated parts of upland Britain with thin soils and sensitive freshwaters (the English Lake District, Wales, and parts of Scotland), eastern Canada and the resistant Canadian Shield, and the Northeast United States. Problems of acidification are much more likely to occur on granite and similarly resistant rocks than on limestones and other carbonates, and it is possible to infer likely sensitivity to freshwater acidification from a knowledge of bedrock geology and the nature of superficial deposits (Figure 3.3).

Soils that (because of underlying bedrock or superficial deposits) are rich in lime have a much stronger resistance to acidification than other soils, and similarly lakes in areas with natural sources of lime have a much greater ability to neutralize or cope with acid deposition than lakes elsewhere.

Figure 3.3 Areas sensitive to freshwater acidification in Europe, based on bedrock geology

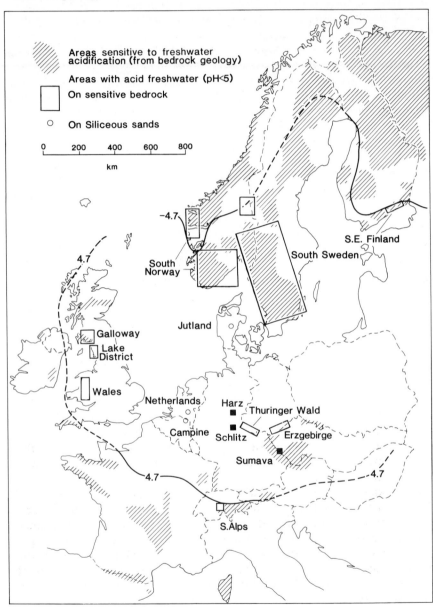

Source: after Wright (1983)

Note: Areas within the pH 4.7 boundary received acid rainfall in the early 1980s. The black boxes show small areas with acid freshwater on sensitive bedrock.

Purposeful man-made additions of buffer capacity

Lime (rich in calcium) is the best-known buffer for acidified soils and waters, which is why it is used extensively in farming, forestry, and gardening.

Lime has been added to lakes, soils, and forests in Scandinavia, West Germany, and Scotland (see chapter 7) in attempts to reduce problems of acidification, but the process is expensive and needs to be repeated regularly to enhance buffer capacity sufficiently to reverse observed biological damage. Moreover, there is mounting concern in Great Britain that the removal of liming subsidies to farmers (in 1976) is leading to reduced addition of lime to many poor soils, which leads to acidification of soils and in turn (through reduced availability of calcium) of lakes and rivers (Taylor 1985).

Inadvertent man-made additions

Acid inputs to lakes and rivers can be effectively neutralized by the increased alkalinity that comes from nutrient enrichment stemming from the (accidental) washout of sewage and agricultural fertilizers. The major man-made source of alkalies is agriculture (Summers 1983), but forest fires also produce alkaline ash, which neutralizes acid soils.

Wind-blown materials

Wind-blown alkaline material can be derived from various sources, including deserts (massive long-distance blows of fine material from the Sahara and Gobi deserts have been recorded; Summers 1983). Other alkaline material can be derived from the wind erosion of top soil, the dispersal of dust from cement works (described in chapter 1), and the emission of particulate pollutants from chimneys. Recall (from chapter 2) how, before the introduction of smoke control Acts in Britain and the United States, alkaline soot particles served to neutralize acids formed in emissions from chimneys.

Most alkaline particles are relatively large (diameter > 10 micrometres), so they quickly settle from the atmosphere – normally within a few hours. Finer particles (2–10 micrometres), on the other hand, have residence times in the atmosphere of up to a week. There is consequently a rapid decline in the concentration of large particles in comparison with the smaller acidic sulphate and nitrate particles, so that downwind the ratio of acidic to alkaline particles increases – i.e. the buffering influence declines (Summers 1983).

Because sensitivity to acidification varies from place to place (in harmony with variations in buffer capacity), it is not possible to arrive at a precise value of precipitation pH below which damage occurs. Consequently the impact of acid deposition can vary considerably, even over short distances. This variability is highlighted, for example, in the English Lake District, which receives among the highest acid fluxes in Britain (Barratt and Irwin 1983). Lakes and vegetation on the Eskdale granites and the Borrowdale volcanics (which offer limited buffering potential) show clear symptoms of acidification, whilst other parts of the Lake District, which receive a similar quality of precipitation, show no signs of damage (Taylor 1985). Some areas – such as Wast Water – are protected by the natural buffering of soils and bedrock, and others with limited natural buffering – such as Esthwaite – are buffered by nutrient enrichment from sewage and agriculture.

UNCERTAINTY AND COMPLEXITY

The acid rain debate is still shrouded in mystery and uncertainty, and this creates two pressing problems – the apparent ease with which ordinary members of the public can become alarmed, and the extreme difficulty of finding widely acceptable solutions to the problem. Kenneth Mellanby has accused journalists of providing 'uninformed nonsense' about acid rain during the early 1980s, and he insists that 'there are still many uncertainties. We simply do not understand the quantitative relationships between emissions and their environmental effects' (1984b: 87). The problem of linearity (see chapter 2) is far from properly resolved.

In particular, the CEGB (1984: 6) in Britain argues that the rainfall acidities being measured in many areas are too high to be accounted for only by gaseous oxides produced from the burning of fossil fuels – unless those gases have been further oxidized (e.g. by ozone) after emission into the atmosphere. This highlights the thorny problem of linearity (see chapter 2), and reinforces the view that it is difficult to attribute relative importance of the many factors involved.

> Is SO_2, for example, responsible [for forest damage] to an extent of 30 percent, 40 percent or 50 percent? The answer must undoubtedly vary, not merely as a function of geographical area – e.g. from one country to another; geographical position (forest on a hillslope or on a valley floor); the environment (nearby aluminium factory and release of fluorine) or the tree's state of health itself determined by the soil; the climate; techniques of forest management, etc; but also as a function of time (seasonal variations).   (Ribault 1984)

The lack of unequivocal evidence linking acid rain with damage to the environment reinforces official views in Britain and the United States that it would be precipitate at this stage to make binding decisions to reduce emissions of SO_2 and NO_x. The apparent uncertainty gives a welcome breathing space to protagonists of more research, and an eagerly seized excuse for inactivity by some governments (see chapters 8–10). This 'need for caution' view is well articulated by the environmental director for America's giant Peabody Coal Company, who argues:

nobody has yet proved a direct relationship between the level of sulphur emissions in the midwest and the amount of acid rain that falls in the northeastern United States and Canada. And until we have this proof, we should go slowly in order to develop the most prudent control scheme. Before being required to . . . reduce emissions, we want assurance that [it] will do some good. (quoted in La Bastille 1981: 665)

The uncertainty reflects the scientific complexity of the problem. There are few really dramatic effects of acidification, and many of the inferred impacts are extremely subtle if not largely invisible (even to the trained eye), at least in the early stages. But by the time visible damage occurs or subtle effects have built up to a detectable (and quantifiable) level, many of the problems are extremely serious and possibly irreversible.

Acidified areas are neither unsightly nor ecologically dead, although many normal species disappear and are replaced by a few acid-tolerant ones. The most serious effects of acid rain reported to date have been large-scale fish kills (for example in Sweden, where over 9,000 previously productive lakes now have depleted fish stocks; Abrahamsen, Stuanes, and Tveite 1983; Swedish Ministry of Agriculture 1984) and widespread forest damage (for example in West Germany, where around one-third of the coniferous forests are troubled by acidification; Tamm and Cowling 1977; Elsworth 1984a; Postel 1984). Moreover, because of variations in buffer capacity, different areas often show different symptoms of acidification, even when exposed to the same levels of acid precipitation.

Complexity also arises because the effects of acid rain are neither simple nor direct. It affects the environment mainly by altering natural equilibria, upsetting the finely tuned natural chemical balances and biological mechanisms on which life depends. It is now believed that acidification *per se* is not the critical issue, at least for wildlife. What is more significant is the chemical chain reactions triggered off by acidification (such as the mobilization of toxic heavy metals – see chapters 4 and 5), which produce the real problems for plants and animals.

Variability of response

Not all effects commonly attributed to acid rain are seen in any one area, and one particular puzzle is why different effects are detected in different areas. In West Germany, for example, tree deaths have caused considerable concern (Straud 1980); in Norway it has been fish deaths (Jensen and Snekvik 1972; Abrahamsen *et al.* 1976; Braekke 1976; Leivestad and Muniz 1976).

The answer lies partly in variations from place to place in the form (wet or dry, also chemical composition) and level of acid deposition (pH and total flux), as outlined in chapter 2. Superimposed on these are variations from place to place in natural buffer capacity, which make some areas more resistant to acidification than others. For example, acidification of freshwater lakes is much less likely in the alpine zone of Europe than in northern Europe because of more favourable geological conditions. Alpine valleys around Lake Maggiore in northern Italy receive acidic rainfall (pH between 4.3 and 4.5), but their underlying geologies (insensitive calcareous rocks) buffer surface waters, which have pH generally in the range 5.0–9.0 (most are between 6.5 and 7.5) (Mosello and Tartari 1983).

Different habitats also respond to given acidity levels in different ways. Thus, for example, freshwater habitats are more readily affected by acidification than soils and vegetation (including trees). Also important is the environment of the area over which acid rains fall, particularly if – as appears to be the case with West Germany's forests – climate is at the extreme of what a given vegetation (in this case coniferous trees) will tolerate. If this is the case, acidification can effectively be the 'straw that broke the camel's back', and even small amounts of acid rain might produce much worse damage than would occur elsewhere (even with much higher acid inputs).

It is also true that visible effects of acidification have been detected in some areas – such as upland Wales (Scottish Wildlife Trust 1985) – where rainfall is not particularly acidic compared with some of the more high-risk areas (such as Sweden and Germany). These areas tend to have high rainfall totals, and the effects are cumulative (reflecting net fluxes of sulphur and nitrogen oxides – see chapter 2). Moreover, the affected areas have a range of environmental conditions that are collectively conducive to acidification. For example, fish deaths in Welsh rivers have been concentrated in areas with hard bedrock, poor soils, and large coniferous plantations.

It is important also to distinguish between effects that are evident close to emission sources (urban and industrial areas) and those that occur long distances downwind. Most of the local problems for humans, vegetation, and buildings are caused by short-range dry deposition of oxides and particles (see chapter 2), and they are often acute and thus relatively easy to detect. Longer-range wet and dry deposition of acids appears to have chronic effects on terrestrial and aquatic ecosystems – trees, soils, lakes and rivers, plants, and animals – and building materials, and these are much more difficult to detect. Strictly speaking the acid rain debate is about the effects of long-range (trans-frontier) acid deposition, but the effects of short-range deposition are also relevant and they must be considered.

Biological acidification

Moreover acid rain is only one component of acidity, and it is necessary – but not always simple – to distinguish between acidification caused by acid deposition and acidification caused by other processes, both natural and man-made.

Natural biological processes have long been known to cause acidification in soils and surface waters. Biological acidification is the result of two processes (Oden 1976: 161–2) – nitrogen metabolism and increasing excess alkalinity in plants – which produce a net effect in which ions are lost from plants in a series of pulses (most notably during spring and autumn). Biological acidification is enhanced when the production of organic matter exceeds decomposition, so that peat bogs and forest peats are normally very acidic. Bogmoss (*Sphagnum* species) secretes acid, heather (*Calluna vulgaris*) increases the acidity of soils and percolating water, and pine forest naturally induces acidification of soils. But the reverse also holds true, and biological acidification decreases when there is a net loss of organic matter (e.g. when crops are harvested on forest and farm land; Bormann and Likens 1979).

Afforestation and acidification

Considerable attention has been focused on the links between acidification and forestry. There are well-founded fears that acid rain is damaging forests in many parts of mainland Europe (see chapter 5), but trees also affect soil acidity.

The acidity of soils is generally lower under deciduous trees than under conifers. Conifers increase soil acidity in two ways – biological acidification within and on the conifers, and chemical enrichment as rainfall passes through the conifer canopy. The acidity of precipitation can be increased by a factor of up to three as it drips from foliage, and by a factor of between eight and ten when it flows down the stems of spruce and pine (Taylor 1985). The washing out of chemical substances from conifer forest litter may increase acidity even further. In contrast, broadleaved trees may serve to reduce the effects of acid deposition by reducing the concentration of strong acids.

Research has shown that commercial conifer plantations can increase the possible damaging effects of acid deposition, so that afforestation can increase the rate of acidification of rivers and lakes. Both chemical and physical effects are involved. The growth of trees removes basic ions from the soil, and this process reduces the ability of the soil to neutralize acidity (i.e. it lowers the soil's buffer capacity). Physical changes are introduced via land drainage (normally by digging dense networks of drainage ditches), which precedes planting. Drainage speeds the removal of water during wet periods, and this reduced contact with the soil decreases the prospect of chemical neutralization of the water through natural buffering. The net effect is to promote acid surges in streams draining coniferous forests, which affect fish and the predatory birds (like dippers) that feed off them.

Thus many forms of forestry management serve to accentuate problems of acidification, and a real paradox emerges. Much marginal land, which is unsuitable for intensive farming use, is very sensitive to acidification (because it has a low buffer capacity, often having acid soils overlying slow weathering rocks) yet it is often regarded as suitable for commercial conifer plantations – which may further increase the problems of acidification. Many parts of upland Britain (including Scotland, Wales, and the English Lake District) and other countries are caught in this self-induced circle of marginality, susceptibility, and accentuation (Harriman and Morrison 1981).

The links between afforestation and acidification are now becoming clear, because the most widespread fish deaths in upland Britain have occurred in areas with hard bedrock, poor soils, and large conifer plantations (*The Times*, 14 January 1983). As a result it is extremely difficult in many areas to separate the possible effects of an increased area planted with conifers from the possible effects of acid deposition (Nilsson, Miller and Miller 1982; Goldsmith and Wood 1983; Binns 1982a).

The time dimension

Another critical uncertainty is why most of the problems attributed to acid rain in Europe and elsewhere are coming to light only now, after a long period during which oxide emissions have been increasing. Sulphur levels in the atmosphere across Europe as a whole rose by around a half between the mid-1950s and the mid-1970s, though levels over Scandinavia and central Europe doubled over the same period (*The Times*, 30 June 1982). Moreover, trends in emission levels have varied considerably from country to country. UK production of SO_2 and NO_x reached a peak in the 1960s, since when it has decreased markedly (partly through the replacement of sulphur-rich coal gas with low-sulphur North Sea gas) (see Figure 2.5) (Central Electricity Generating Board 1984; HMSO 1984). The puzzle is made even more complex by the fact that, although the pH of rainfall over these areas is presently low (i.e. rainfall is very acidic), there is no strong evidence to indicate that the acidity has been increasing steadily in recent decades (see chapter 2).

The evidence suggests that acid rain has effects that are indirect, cumulative, and difficult to detect with precision. There has been a time-lag between the period of peak SO_2 emissions and likely peak of acid deposition (both in the 1950s and 1960s in Europe – Wright 1983: 142), and the period of maximum damage observed, or at least reported (the 1980s and probably longer). The lag may reflect two trends superimposed on one another. One is the gradual rundown of buffer capacity over time, so that an acid 'war of attrition' slowly but progressively grinds the environmental systems into submission and eventual defeat. The other is the 'domino effect': the progressive and cumulative breakdown through time of natural environmental systems (chemical balances and biological mechanisms), so that as time progresses more and more elements in the system (or dominoes in a precariously balanced stack) are disturbed, especially as ecologically significant thresholds or critical break-points are reached.

Every cloud has a silver lining

It is perhaps appropriate to end this chapter on a positive note. Acid rain has had a bad press in recent years, and the implication is that it always spells trouble for the environment and for humans. Whilst this is doubtless true overall, not all news of acid rain is bad news and some beneficial effects have been reported. The silver lining in some acid rain clouds is the prospect of gaining free nutrients.

The sulphur and nitrogen oxides in acid precipitation can provide valuable nutrients that act as free fertilizers and stimulate plant growth. It has been known for over thirty years, for example, that atmospheric SO_2 could be usefully absorbed directly by the shoots and leaves of plants. Studies carried out at Britain's Grassland Research Institute have found that low concentrations of atmospheric SO_2 can increase the growth of the perennial ryegrass (*Lolium perenne*), a common field grass on some soils (Cowling, Jones, and Lockyer 1973; *The Times*, 2 July 1973).

More recently, studies off the coast of North Carolina have found that nitrogen-enriched acid rains appear to enhance the growth of phytoplankton – the microscopic marine plants that larger fish feed on (Paerl 1985; *The Times*, 2 July 1985). This might significantly benefit commercial fisheries in shallow coastal waters close to urban and industrial emission sources of sulphur and nitrogen oxides. Such conditions are typical of the United States, the United Kingdom, the Baltic region, and the western Pacific waters of Japan, Korea, and China, and the potential economic benefits – as yet unassessed – should not be overlooked when evaluating acid rain in global terms.

Other examples of the potential benefits of nutrient enrichment by acid rain are starting to emerge from many areas. However, the evidence available to date indicates that such benefits tend to be localized. Overall, as yet, the bad news (summarized in chapters 4, 5, and 6) far outweighs the good.

CONCLUSION

Over the last decade or so, evidence of environmental damage from acidification, which many scientists attribute either directly or indirectly to acid rain, has started to appear in many developed countries. The damage attributed to acid rain includes declining fish populations, leaching of acid minerals from soils into surface waters, forest decline, and increased plant disease (Cowling 1982).

The widely documented evidence of fish and tree deaths, in particular, has been used extensively in political debate and in the media to highlight the urgency of the acid rain problem and to stress the need to embark on long-term programmes of remedial action (particularly the reduction of SO_2 emissions from power station chimneys) without further delay. In practice, however, it is extremely difficult to establish definite direct links between SO_2 emissions, rainfall acidity, and observed effects on the environment. This does *not* necessarily mean that these three components of the debate are not linked causally. What it does mean is that unequivocal evidence, strong enough to be entirely convincing to politicians and the electricity-generating authorities (CEGB in Britain) and companies (in the United States), is rather difficult to come by.

In the United Kingdom, the Parliamentary Select Committee that investigated acid rain in 1984 (see chapter 10) blamed the CEGB for most of the observed damage to trees, lakes, and buildings, whilst the government took refuge in the call for more scientific evidence before money was spent on reducing emissions. But the Members of Parliament who sat on the Committee were 'convinced that scientific understanding is sufficiently advanced to demand action now', and they concluded that 'while a direct and proportional link between sulphur emissions and environmental damage is not scientifically proven, the case against SO_2 is telling enough and the damage severe enough that action must be taken before all the evidence is in' (Walgate 1984: 94).

A similar conclusion was drawn in 1984 by a group of environmental consultants called in by the European Community to review the acid rain problem in Europe:

> it has not been unequivocally established that [these] environmental impacts are caused by acid pollutant emissions . . . nevertheless circumstantial evidence would suggest that acid emissions and their subsequent chemical transformation and precipitation are at least a partial contributory cause to the observed effects and may be giving rise to as yet unidentified impacts, some of which could be irreversible. (Environmental Resources Ltd 1984: 20)

In the next three chapters we shall have the opportunity to examine the main lines of evidence – circumstantial or otherwise – that have been used to link environmental damage with acid deposition, and to evaluate the claims made by many environmental groups that politicians in some countries (like Britain and the United States) have viewed the acid rain problem with contemptuous indifference.

4 EFFECTS ON SURFACE WATERS
'Down by the riverside . . .'

One may not doubt that, somehow, good
Shall come of water and of mud;
And, sure, the reverent eye must see
A purpose in liquidity

(Rupert Brooke, *Heaven*)

As far as we are concerned in this book, the 'purpose in liquidity' is the
rapid transfer of dissolved acids through the environment. Both dry and
wet deposition contribute acid substances to natural biogeochemical
cycles, which can show clear symptoms of environmental damage from
effective 'overloading'. These natural cycles also move nutrients through
ecosystems, and so they widen the area over which ecological damage from
acidification can become evident. Thus acidification can – and indeed does
– create problems in habitats some way downstream from those areas that
receive the acid deposition (which are themselves long distances downwind
of emission sources), as well as in the immediate vicinity of the deposition.

The aim in this chapter is to review some of the evidence that might link
ecological changes in aquatic ecosystems with acidification (whether this be
induced by acid rain or natural processes). Ecological damage to lakes and
rivers has been reported in many areas since the mid 1970s (Hagen and
Langeland 1973; Almer *et al.* 1974; Beamish *et al.* 1975; Abrahamsen *et al.*
1976; Braekke 1976; Department of Environment 1976; Drablos and
Tollan 1980; Fromm 1980; Muniz and Leivestad 1980a, b; Overrein, Seip
and Tollan 1981; Mosello and Tartari 1983; Vangenechten 1983; Wright
1983; Environmental Resources Ltd 1984; Royal Society 1984; Swedish
Ministry of Agriculture 1984), and the observed changes follow a series of
quite clearly defined trends and patterns. The damage is random neither
geographically nor ecologically, and this lends substance to tne claims that
SO_2 and NO_x emissions appear to be the ultimate source (however
modified in transit through the atmosphere – see chapters 2 and 3) of the
acid precipitation that appears to cause fish deaths and biological changes
in freshwater habitats.

ACIDIFICATION OF SURFACE WATERS

In many areas (particularly in Scandinavia and the Northeast United States) it has been suggested that increased acidity (declining pH) of precipitation has been paralleled by a decrease in the pH of stream and lake waters. Surface waters offer much less prospect of buffering acid inputs than do soils and plants. Consequently streams and lakes normally show clear signs of acidification some time before it becomes apparent in soils, plants, and forests.

The most dramatic symptoms are the decline and possible loss of populations of fish and other species from acidified lakes, rivers, and reservoirs (Likens *et al.* 1979), and it comes as no surprise that such evidence has been widely used by conservationists in their campaign to reduce SO_2 and NO_x emissions from power stations, houses, vehicles, and industry. In southern Norway, for example, fishery inspectors at the turn of the century noted fish kills of salmon possibly related to acid waters, and brown trout began to disappear from many mountain lakes during the 1920s and 1930s. By the 1950s, barren lakes were being reported from many parts of southern Norway, and commercial fishery catches were declining in parts of Sweden, especially along the west coast (Wright 1983: 137).

But fish deaths are merely the tip of the acid iceberg, because acidification affects surface waters in a variety of ways that are not always visible or easily detected.

It is difficult to specify at what precise point a lake or river becomes 'acidified', because surface waters in some (well-buffered) areas can tolerate much higher levels of acidity than others. The definition should be based on biological criteria (e.g. damage to fish), but this is closely correlated with a simple chemical definition. It is widely agreed that, once the pH of surface waters falls below 6.0, chemical and biological conditions change so significantly that for all intents and purposes the water can be regarded as 'acidified'. As pH falls even lower, the acidification intensifies and the problems magnify and diversify.

EVIDENCE OF RECENT CHANGES

Acidified lakes (pH less than 6.0) have now been reported from many areas. The worst-hit areas up to now have been southern Norway (Figure 4.1a) and Sweden (Figure 4.1b), the Adirondack Mountains in New York state (USA), south-east Ontario (Figure 4.1c), and other parts of eastern Canada (Wright *et al.* 1980a; La Bastille 1981). But acid lakes have also been found in Belgium, Denmark, West Germany, and the Netherlands (Whelpdale 1983: 78), and scientists expect symptoms of acidification to appear in more and more countries if recent increases in rainfall acidity continue unchecked.

Great Britain has not escaped untouched. Since 1982 some lakes in Galloway in south-west Scotland, and some rivers in the English Lake

Figure 4.1 Distribution of acidified lakes in southern Norway, western Sweden, Ontario, and alpine Italy

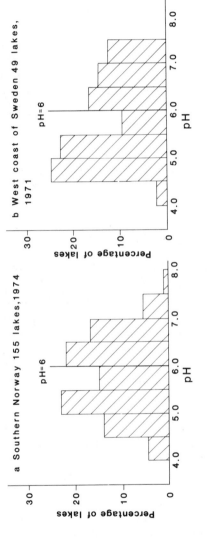

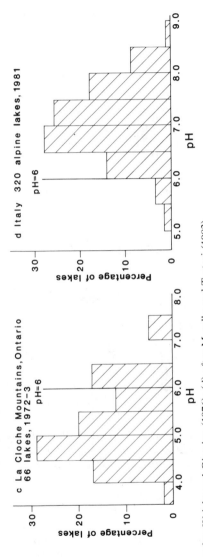

Sources: a–c after Wright and Gjessing (1976); (d) after Mosello and Tartari (1983)
Note: The data are plotted as frequency distributions of the pH in lakes, expressed as a percentage of lakes falling in each successive pH class. pH below 6.0 is taken to be 'acidified'.

District and in central Wales have been added to the casualty list (Wright *et al*. 1980b; Sutcliffe *et al*. 1982; Battarbee 1984; Scottish Wildlife Trust 1985; Sutcliffe and Carrick 1986). Most of these areas have granite or other highly siliceous bedrock, soft and poorly buffered surface waters, and highly acid precipitation (average pH below 4.5).

Lakes in well-buffered areas, such as the alpine lakes of northern Italy (Figure 4.1d) are generally much less acidic than those in poorly buffered areas (contrast Figure 4.1d with 4.1a–c). The acid lakes in Belgium and the Netherlands are related more to the presence of thick deposits of siliceous sands than to acid precipitation (Wright 1983: 138).

It is difficult to establish how much the acidity of surface waters has changed through time in most areas, because it is only recently that reliable and regular measurements have been taken. What evidence we do have shows that many of today's acidified lakes were not so even in the recent past. The pH of 187 lakes in southern Norway was measured at various dates between 1923–49 and the 1970s, and the results (Figure 4.2a) show a fall in pH (in the order of 1 pH unit, i.e. a ten-fold increase in acidity) over that period (Wright 1977). Similar measurements indicate a marked fall in pH in 51 lakes in southern Sweden between 1935 and 1971 (Figure 4.2b), and in lakes in New York state's Adirondack Mountains between the 1930s and 1975 (Figure 4.2c). Observations of the changing water chemistry of George Lake in Canada (Beamish *et al*. 1975) show that pH fell on average around 0.13 units per year between 1961 and 1975. This represents a total fall of around 1.7, or a fifty-fold increase in acidity over the fourteen-year period. A less dramatic but none the less important drop in pH has been noted for the River Klaralven in Sweden, from 6.9 in 1965 to 6.5 in 1974 (Oden 1976: 154–5).

However, whilst such water chemistry evidence provides a clear picture of changing levels of pH in surface waters, it does mask considerable variability in acid levels over short time-periods which might be of greater biological significance. For example, short bursts of acidity – perhaps during spring snow melt or after prolonged drought (see chapter 2) – can cause wholesale biological disturbance but escape detection in the general figures on long-term changes in surface water pH.

Biological records can also be used to confirm how recent most surface water acidification has been. Over 20 per cent of the lakes in southern Norway that are now barren lost their fish during the 1970s, and in over half of the 700 lakes with sparse fish populations the decline has been very recent (Wright 1983: 140). Between 1929 and 1937, only a few lakes in the Adirondack Mountains in the Northeast United States had a pH below 5.0, but by the mid 1970s almost half of them had (see Figure 4.2c), and by then 90 per cent of the lakes had lost their fish stocks (Schofield 1976).

The evidence indicates a progressive decline in water quality (i.e. a fall in pH) in all areas that receive high inputs of acid substances (whether from natural or man-made sources). If recent trends continue unchecked, therefore, we should expect to find progressively more and more streams

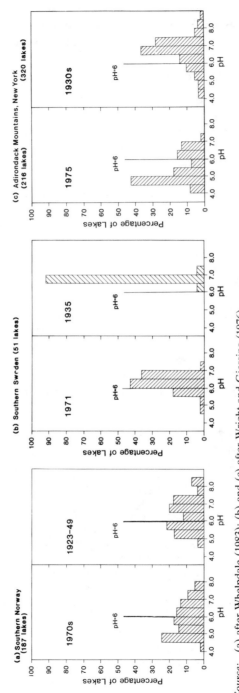

Figure 4.2 Increasing acidity of lakes through time in southern Norway, southern Sweden, and the Adirondack Mountains (Northeast United States)

Source: (a) after Whelpdale (1983); (b) and (c) after Wright and Gjessing (1976)

and lakes in many parts of the United States, eastern Canada, Scandinavia, and north-west Europe (including Great Britain) showing signs of acute acidification – including decline and loss of fish populations.

ACIDIFICATION PROCESSES

Surface waters, which include streams and rivers, lakes and reservoirs, rarely have 'pure' water. Pure water has a pH of 7.0 (see Figure 2.1), and only lakes and rivers in unpolluted areas with alkaline bedrock and deposits – such as the western Pacific region (Table 4.1) – have water of anything approaching this level of 'purity'. Most surface waters in other areas are contaminated with materials derived from natural geological and biological processes that take place within their catchment areas, and from environmental pollution. The level and character of the contamination varies considerably from place to place, and it also varies through time at a given place (Faust and Hunter 1976; Alabaster and Lloyd 1980; Haines 1981; Walling and Webb 1981). Consequently it is often difficult to establish exactly how 'natural' the water chemistry at a particular site is, particularly (as is more often than not the case) in the absence of long-term monitoring.

Acid substances exist naturally in all water bodies, and acid precipitation may be relatively minor in many areas. Studies of the chemical composition of seventy-two lakes in Galloway, south-west Scotland (Wright *et al.* 1980b), indicate three major source areas for ions found in the lakes (Table 4.2). The composition of atmospheric chemistry (reflecting both natural and man-made sources) varies from place to place (see

Table 4.1   *Acidity of river and lake waters in countries of the western Pacific region*

Country	River or lake	pH
Philippines	La Mesa Reservoir	7.1–8.2
	Laguna Lake	6.9–8.9
	Pampanga River	7.7–8.3
	Cagayan River	7.2–8.4
Australia	Murray River	7.7–8.4
	Mount Bolo Reservoir	6.2–8.4
	Murrumbidgee River	7.1
	Darling River	7.6–8.8
	Lake Burragorang	7.4–7.7
Malaysia	Kelang River	6.1–7.5
	Gombak River	6.7–7.5
	Linggi River	6.2–7.3
	Sekudui River	5.9–6.6
	Muda River	6.1–7.4
	Kinta River	6.7–7.5
	Waimann River	6.8–7.2

Source: Okita (1983)

Table 4.2   *Source areas of ions in lakes in Galloway, south-west Scotland*

Source	Major ions contributed
atmospheric inputs of seawater	sodium (Na^+), potassium (K^+), magnesium (Mg^{++}), chloride (Cl^{--}), sulphate (SO_4^{--})
atmospheric inputs of pollutants	hydrogen (H^+), ammonium (NH_4^-), nitrate (NO_3^-), sulphate SO_4^{--})
terrestrial inputs of weathering products	calcium (Ca^{++}), magnesium (Mg^{++}), potassium (K^+), aluminium (Al^{+++}), bicarbonate (HCO_3^-)

Source: after Wright *et al.* (1980b)

chapter 2), and the relative importance of the three sources will also vary from place to place.

Acid precipitation can be buffered (or effectively neutralized – see chapter 3) en route to the river or lake. Thus the impact of acid rain is *not* simply a product of the pH of the water, because increased acidity of infiltrating water may lead to increased leaching of cations from the soil (as a result of cation exchange), and may trigger off accelerated weathering of silicate minerals and the solution of heavy metals that have accumulated in the soil (see chapter 5). Natural surface waters are generally buffered by soluble substances (phosphate (PO_4^{---}), organic acids, amino acids), colloidal (invisible to the naked eye, but visible under powerful microscopes) organic matter, and bicarbonate. Rivers with low buffering capacities are very sensitive to the addition of acids, especially in the range pH 6.5–4.5. Very small additions of acid to such lakes or rivers may cause a rapid drop from pH 6.5 to pH 4.5 (Oden 1976: 155), which is generally below the threshold of acidity for biological damage (see next section).

The scale and speed of impact of acid precipitation on a water body depends partly on the processes involved. Wet deposition (dissolved acids) and dry deposition (gases and aerosols) on to the water surface provide an instantaneous acid input. Material that is dry deposited on vegetation overhanging the water surface and on soils and vegetation in tributary areas (e.g. during summer drought) is subsequently activated by dissolution in the first rains, which wash it into the water body. Massive acid surges over a limited period often accompany the end of prolonged drought or spring snow melt (see pages 77–82).

Direct inputs of acid via precipitation normally play a relatively minor role in the overall acidification of water bodies, even in areas that receive large amounts of low pH rainfall. The most important inputs normally come from the soils, vegetation, and rocks in tributary areas, so that the impact of acidification on water bodies is closely related to its impacts in surrounding areas. Natural processes of chemical weathering or breakdown release acid products from soils, deposits, and bedrock, and these acids are dissolved and leached in the natural waters that drain into lakes and streams. Lakes and rivers thus have a natural background acidity that reflects soils and rocks in their tributary areas. When soils are disturbed by acid rain, soil water chemistry is changed in three ways: its acidity increases, its dissolved nutrient content alters, and heavy metals (including aluminium) that are naturally present are mobilized (i.e. they can be leached out) (Braekke 1976; Department of Environment 1976; Likens *et al.* 1977; Cronan and Schofield 1979; Drablos and Tollan 1980; Hutchinson and Havas 1980; La Bastille 1981). This soil water, with its altered chemical composition, eventually finds its way into streams and lakes and in turn alters their chemistry.

Natural sources and complications

Natural acidification processes within soils and vegetation (Oden 1976; Likens *et al.* 1977; Krug and Frink 1983; Tabatari 1985) complicate the picture in at least two ways. First, some ecological changes are promoted by natural acidification (see chapter 3), so that not all acid damage to the environment is derived from air pollution. Secondly, where both natural acidification and acid deposition occur together, it is extremely difficult to establish which problems are caused by which source of acids. As a result, reduction of SO_2 and NO_x emissions may bring only partial relief in acid-damaged areas (the linearity question again). Moreover, such reductions in emissions might not prevent wholesale ecological damage arising in sensitive habitats that are still being subjected to natural acidification.

A number of natural biological processes produce acids that are washed into lakes and rivers and in turn lower their pH. A change in land use (e.g. moorland to farming) or management (e.g. adoption of intensive chemical farming practices) within the tributary area of a lake, reservoir, or river can seriously alter the rates of natural production of acids, so that acidification can occur in the absence of acid rain. For example, at least part of the increased acidity recently observed in stream-draining coniferous forests in parts of Scotland may be attributed to natural acidification within the forest ecosystem (Harriman and Morrison 1981, 1982). The solution to acidification damage might thus in some cases rest partly in forest management practices and afforestation policies, as well as in reducing SO_2 and NO_x emissions.

EFFECTS OF ACIDIFICATION ON SURFACE WATERS

The direct effect of acid precipitation is to change the chemistry of surface waters. The indirect effect, which occurs via altered water chemistry, is to change the suitability of the aquatic habitat for different plant and animal species. Conservationists are quite understandably more concerned about the indirect effects, which form a central ingredient in the acid rain debate.

Surface waters that have become acidic are described as 'oligotrophic'. The biochemical processes involved (oligotrophication) represent the reverse of nutrient enrichment processes (eutrophication), which are common in intensively farmed areas and which arise from the washing out of large quantities of chemical and natural organic fertilizers from fields (Mellanby 1972; Gower 1980: 147–53; Park 1981: 203–4).

The ecology of a lake or river alters when it becomes acidified, but acid lakes are not necessarily biologically 'dead'. Many plant and animal species, particularly the widespread ones, cannot tolerate acid water and so population levels decline and ultimately the species disappears. The ecological community in turn becomes dominated by a small number of acid-tolerant species, which survive in or colonize the acidified surface waters.

Tolerance, stress, and thresholds

Any species of plant or animal has a 'tolerance range', or natural inbuilt ability to tolerate a range of environmental conditions (Fromm 1980; Park 1981: 94–9). There is an optimum environment for a given species, to which it is best adapted and in which it will grow best, survive longest, and reproduce most successfully. All environments face regular cycles of seasonal change (e.g. in temperature and sunlight) and the species within them are biologically adapted (through physiology or behaviour) to cope with these. But longer-term environmental change (e.g. acid rain) can bring sub-optimal conditions that create stresses for the species concerned.

The tolerance range of a particular species may accommodate minor environmental changes, so that the population survives but adjusts. Adjustment can take various forms: productivity may decline (i.e. individuals grow more slowly), reproduction rates may decline (fewer young are born), population structure may change (older individuals may not be able to survive the change in conditions, so they die prematurely). Larger changes, especially those that cross critical thresholds of tolerance, may produce acute physiological stress for the individual plants or animals involved. Such changes can also increase the susceptibility of those individuals to minor changes in the environment (such as drought), often culminating in death.

The threshold of rainfall acidity (weighted average) is believed to be between pH 4.6 and 4.7; below this level acidification of unbuffered or poorly buffered surface waters is likely to occur (Wright 1983: 141), and their pH falls below the critical level of around 6.0.

Tolerance thresholds and ecological changes

Figure 4.3 summarizes the general pattern of biological changes associated with acidification of surface waters, based on the types of population changes in acidified lakes and rivers that have been reported in Norway, Sweden, Canada, the United States, and more recently Great Britain.

The critical pH level for most aquatic species is 6.0. Below this the ecology of the lake or river changes significantly – the number and variety

Figure 4.3   Changing tolerance of freshwater species as water pH falls

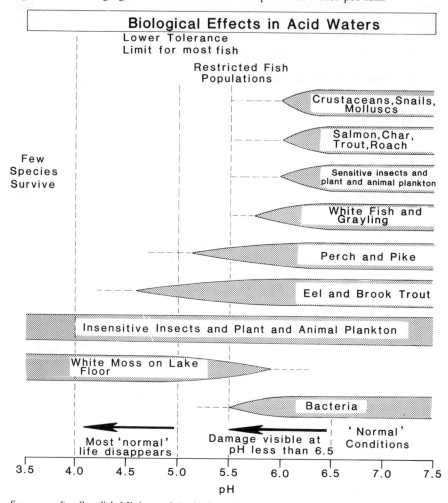

Source: after Swedish Ministry of Agriculture (1984) and McCormick (1985)
Note: Most fish populations decline below pH 5.5, few survive at pH 5.0, below pH 4.0
   there are few species of anything left

of species decline, and in particular there are smaller and less diverse populations of algae, zooplankton, aquatic insects, insect larvae, and sensitive fish (such as salmon, roach and minnow) (McCormick 1985: 34; Berlekom 1985c). Below pH 5.5, the number of snails and phytoplankton falls; below pH 5.2, snails disappear; below pH 5.0, zooplankton disappear; when pH falls below 4.0, stocks of all fish species decline rapidly (because embryos fail to mature at this level of acidity) (Carrick 1979; Bryce-Smith 1985).

Different species have different tolerance ranges for given aspects of the environment (e.g. pH). Thus environmental stress promotes changes in species composition, so that some species (now beyond their optimum) decline or disappear whilst others (now approaching their optimum) thrive and dominate. There are normally far fewer of the latter, so that ecological diversity (the number of different species present) declines with acidification.

Sensitivity to environmental change is a reflection of breadth of tolerance range, and it varies through the lifecycle of an individual as well as from species to species. Young and old individuals of a population are often the most sensitive, and so they disappear more readily than the middle age-groups. Fish and amphibians, for example, are especially sensitive to acidity during their early embryonic stages. Population structure within a species can thus change as a result of environmental stress.

An interesting example of acidic lakes is offered in the so-called 'smoking hills' of the Canadian arctic (Havas and Hutchinson 1983; *The Times*, 8 January 1983), where some freshwater ponds are naturally alkaline yet others (which are fumigated by sulphur-rich winds blowing from downwind industrial areas) are acidic. The beds of the acidic ponds are covered by dense mats of mosses, and none of the eleven species of microscopic animals and two insect larvae present in the alkaline ponds survive in the acidic ones. Only species tolerant of high levels of acidity and high concentrations of toxic metals (such as aluminium, iron, and manganese) survive in the acidic ponds.

The community of micro-organisms, both plants (phytoplankton) and animals (zooplankton), is different in acid waters than in more normal waters: fewer individuals and fewer species are present. Most of the green algae and diatoms normally present in freshwater lakes disappear when pH falls below 5.8 (Almer *et al.* 1974). A reduction in the number of small plants allows light to penetrate further into the water, so that acidified lakes are generally crystal clear and bluish in appearance.

Diatoms are small siliceous phytoplankton, and different diatom species can tolerate different levels of acidity. Consequently studies of the changing diatom populations preserved in fossilized forms at different depths within the sediments on the bed of acidified lakes can reveal a great deal about the probable changes through time in acidity of the overlying lake water (Davis and Berge 1980; Battarbee *et al.* 1985b). The diatom evidence suggests that the pH of lakes in the Galloway region of Scotland, for example, started to fall around 1830 and has fallen considerably over the present century (Figure 4.4).

Larger aquatic plants (macrophytes) often decline, but acid-tolerant white mosses (*Sphagnum*) colonize acid lake beds. *Sphagnum* mosses and filamentous algae grow very fast and very large in acid waters (with pH below about 5.5; Figure 4.3). They can form impenetrable mats that seal off oxygen and slow down the decay of litter on the lake floors (La Bastille 1981: 672; Pearce 1982a). Leaves remain largely unrotted on the lake beds,

Figure 4.4   Recent fall in pH of Loch Enoch, south-west Scotland, inferred from diatom species in the lake floor sediments

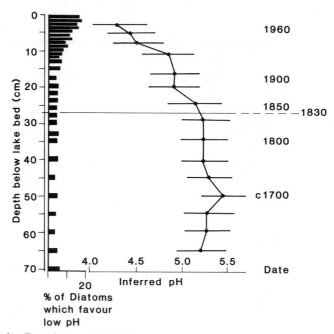

Source: after Battarbee *et al.* (1985b)

and the enhanced visibility through the clear water allows them to be seen more clearly. In acidified lakes the rate of decomposition of organic matter often slows down because the fungi and bacteria that play key roles in this decomposition are not tolerant of acidic conditions – they tend to disappear when pH falls below about 5.5 (Figure 4.3). Because the organic debris decomposes more slowly than normal, the mass of vegetation on the acidified lake bed can trap valuable nutrients rather than releasing them back to the lake ecosystem.

Many species of amphibians – including frogs, toads, and salamanders – are particularly sensitive to small changes in water acidity, and low pH greatly stunts their development (Whelpdale 1983: 79; Berlekom 1985b). The numbers of frogs and toads have fallen in some Swedish lakes when acidity increased and *Sphagnum* moss spread over the lake bed. For example, the gradual but progressive decline in the frog population of Lake Tranvatten in Sweden is a sad tale of local extinction brought about by acid stress (Table 4.3). But acid-tolerant amphibians, such as the smooth newt (*Triturus vulgaris*), benefit from such changes because they can survive the falling pH and expand into areas previously occupied by frogs and toads (Dudley, Barrett, and Baldock 1985).

Table 4.3 *Declining frog population in Lake Tranvatten, Sweden, 1973–9*

Date	Effect on frogs
1973	all eggs died
1974	hatching partly successful, but only 25 tadpoles were found from 200 egg clusters, and all later disappeared
1975	20 per cent hatching success; all died in May
1976	only a few tadpoles appeared; all soon died
1977	all eggs died before tadpoles were hatched
1978	only 4 egg clumps were found; all died within a few days of hatching
1979	frog now extinct in the lake

Source: Hagstrom (1980), cited in Dudley, Barrett and Baldock (1985)

Acid stress can lead to a decline in the diversity of aquatic fauna because impacts can be transmitted along food webs by population changes. Thus the loss of micro-organisms and larger plants decreases the availability of food for grazing animals (herbivores), so fewer of them survive. For example, a decrease in bottom-dwelling animals (benthos) can lead to a decline in the number of species of flies, mosquitoes, craneflies, and midges (Diptera) and mayflies (Ephemeroptera) present in acidified waters (Likens 1985). In turn this declining population of grazers decreases the availability of food for carnivores (which eat herbivores), so fewer of them can survive. Predatory birds, such as flycatchers, which nest on the shores of acidic lakes, eat the fish and end up with high concentrations of aluminium in their bodies. In turn, they produce eggs with soft or no shells, and so fewer young are hatched successfully (May 1982: 10). Recent studies in north and mid-Wales have shown a decline in the breeding distribution of dippers (*Cinclus cinclus*) living close to acidic streams, brought about by a reduced food supply of the macro-invertebrates (such as caddis larvae and mayfly nymphs) on which dippers normally feed (*The Times*, 7 August 1985; Berlekom 1985a). The size of dipper territories in southern Scotland is also known to have increased since the mid-1960s, and their food supplies have become more scarce through water pollution (Taylor 1985).

The complex food webs typical of unpolluted water bodies are progressively simplified as acidification alters species structures in the polluted lakes and rivers. But fewer predatory animals means greater opportunity for populations of acid-tolerant species of insects and plants not only to survive (Figure 4.3) but possibly even to expand through lack of competition. Waterbugs (Hemiptera and Heteroptera) and waterboatmen (Corixidae) are acid tolerant because they are physiologically well suited to acid stress. They can survive down to a pH of at least 3.4, and they often dominate acidified lakes (Vangenechten 1983).

Most aquatic animals are more seriously affected by changing water chemistry than by changing food supplies. The pattern is often similar from place to place: rates of survival and reproduction in acid-intolerant species fall, population size decreases (there are fewer individuals of a species), species diversity declines (there are fewer species, e.g. of crustacean zooplankton – see Figure 4.5b), and the total number of organisms present falls. Reductions in both the number and size of individuals within a population in acid lakes also mean a reduction in biomass (weight of living organic matter). A reduced biomass of benthic invertebrates in acidified lakes is clear in Figure 4.5a.

Cutbacks in the amount of biological activity in acidified lakes in turn produce changes in the water: dissolved oxygen levels decline and the water becomes clearer (thus transmitting sunlight to even greater depths).

Fish deaths

Most biological concern has centred on severe declines and losses of fish populations in many acidified lakes and rivers, especially in Eastern North America and in Scandinavia. Ontario in Canada has over 100 acidified lakes that are now completely fishless (Whelpdale 1983: 79), and over the last two decades lake trout, wall-eye, burbot, and small-mouth bass have disappeared from George Lake in Canada (Beamish *et al.* 1975). The record of fish decline in Lumsden Lake, Ontario, since the 1950s (Table 4.4) shows a pattern typical of many acidified lakes.

Many of the high-level (above 610 m) lakes in the Adirondack Mountains in the Northeast United States with pH below 6.0 have no fish populations left (Figure 4.5c). In southern Norway, fish populations have disappeared from an area of 13,000 sq. km and fish in a further 20,000 sq.km are showing signs of stress (Jensen and Snekvik 1972). Roach, arctic char, trout, and perch have disappeared from many acidified lakes in Sweden (Almer *et al.* 1974).

Table 4.4   *Record of fish decline in Lumsden Lake, Ontario*

Date	Species information
1950s	eight fish species present in the lake
1960	last report of yellow perch (*Perca flavescens*)
1960	last report of burbot (*Lota lota*)
1961	lake pH measured at 6.8
1960–65	failure of sport fishery
1967	last capture of lake trout (*Salvelinus namaycush*)
1967	last capture of slimy sculpin (*Cottus cognatus*)
1968	white sucker (*Catostomus commersoni*) suddenly rare
1969	last capture of trout-perch (*Percopsis omiscomaycus*)
1969	last capture of lake herring (*Coregonus artedii*)
1969	last capture of white sucker
1970	last capture of lake chub (*Couesius plumbeus*)
1971	lake pH measured at 4.4; one fish species remains

Source: Beamish and Harvey (1972)

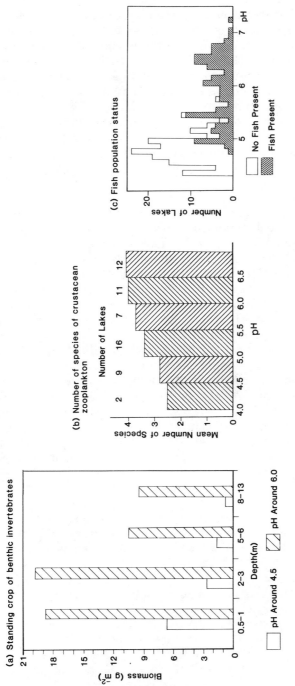

Figure 4.5 Some biological changes in freshwater lakes at low pH.

Sources: (a) after Hendrey *et al.* (1976); (b) after Leivestad *et al.* (1976); (c) after Schofield (1976)

Although long-term progressive changes in the acidity of surface water produce changes in fish populations, the most dramatic mass fish deaths seem to occur in lakes and rivers during or shortly after acid surges induced by spring snow melt (Leivestad and Muniz 1976; Crawshaw 1986).

Complete loss of fish is an extreme position, at the end of a spectrum of ecological stress. Many more lakes and rivers have lost some (the acid-intolerant ones) but not all of their native fish populations. Surviving populations are often depleted in number and diversity, show symptoms of stress (e.g. they are very susceptible to disease), and may have altered age distributions (with perhaps few young and old members). Surveys of the status of fish in 1,679 lakes in southern Norway (Figure 4.6) indicate that once pH falls below 6.0 healthy populations are progressively replaced by sparse and totally depleted populations.

In Scandinavia the decline in fish populations has been particularly severe. Fish have disappeared from nearly one-fifth of the 5,000 freshwater lakes in southern Norway (Tolba 1983: 116), and many of the previously productive salmon rivers in the south and south-west no longer have viable fish populations (Dr Rosseland quoted in *The Times*, 13 October 1983), whereas other Norwegian rivers have not been so affected. Brown trout have disappeared from over 1,400 lakes in southern Norway since 1940 (Figure 4.7). Similar declines are recorded for around 9,000 lakes in southern Sweden, and salmon stocks have been seriously reduced in many rivers along the west coast of Sweden (Wright 1983: 140). Up to 20,000 of Sweden's 90,000 freshwater lakes are now believed to be affected by increased acidity (5,000 of them badly); some have no fish left at all and many have few survivors (Dr Bengtsson quoted in *The Times*, 13 October 1983).

The decline of fish populations is serious not only from an ecological point of view – it can also have important sporting and commercial implications. Several rivers in Nova Scotia, Canada, are now too acidic for successful reproduction of Atlantic salmon (Whelpdale 1983: 79). Between 1970 and 1978 sports fishing catch rates in Quebec's Laurentides Park fell by around one-third, and this was blamed on acidification of the lakes (Bobee and Lachance 1983: 160). However, at least in the early stages of acidification, appearances can be deceptive (May 1982: 10). Old fish can survive in acid waters, and they are able to grow big because competition for food is decreased. Thus anglers might – and do – believe that fish populations in their acidifying lakes are improving, whereas in reality this short-term gain signals imminent population decline through loss of reproductive capacity.

Most of the reported damage to fish populations has come from Scandinavia and North America, and up to the late 1970s Britain appeared to have escaped almost untouched. But recent studies of Galloway in Scotland reveal that fish are now absent from several acidified lakes, and trout and pike populations in many surrounding lakes are much depleted (Wright *et al.* 1980b). Loch Enoch now has no fish, and Loch Grannoch's

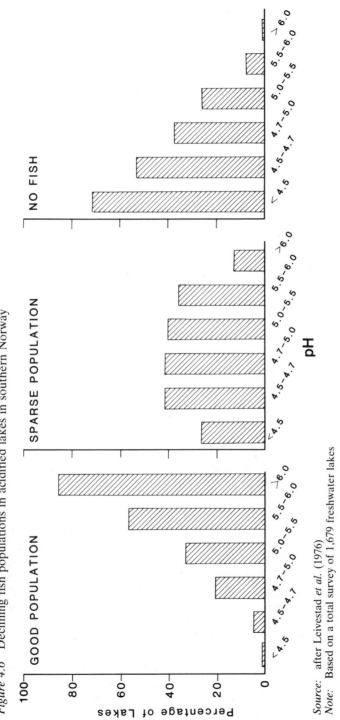

Figure 4.6 Declining fish populations in acidified lakes in southern Norway

Source: after Leivestad *et al.* (1976)
Note: Based on a total survey of 1,679 freshwater lakes

Figure 4.7 Loss of brown trout population from rivers in southern Norway

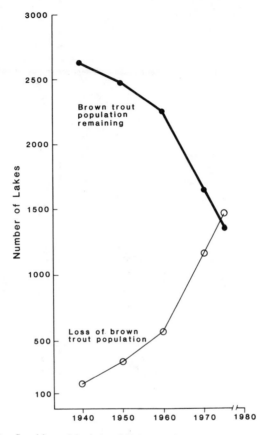

Source: after Sevaldrup, Muniz, and Kalvenes (1980)

long history of successful fishing has drawn to a close (*The Times*, 13 October 1983). Fish farms are also being threatened in Britain. Over 20,000 trout have died in the acidified waters of one fish farm in Galloway, owned – ironically – by the chairman of the EEC Advisory Committee on Fish Farming. He noted in October 1983 that his was 'one of the first [fish farms] to show signs of it [acid rain] in Britain. We're seeing the first signs here that they saw in Norway twenty years ago' (Graham Gordon quoted in *The Times*, 13 October 1983).

Acidification also appears to be affecting rivers and lakes in other parts of central Scotland, although the contribution of natural acidification brought about via intensive reforestation practices remains as yet unclear (Harriman and Morrison 1982). The effects of acid rain on fish populations depend on the pH of the precipitation, but also on the underlying geology (thus buffer capacity of the area) and the extent to which afforestation practices modify the nutrient balance of the area (see chapter 5).

Signs of fish deaths are also emerging some 30 km south-east of Galloway, across the English border in Cumbria. Kills of adult sea trout and salmon have been reported in the River Esk, which drains the western Lake District (Oglesby 1981; Sutcliffe 1983; Sutcliffe and Carrick 1986; Crawshaw 1986). Wales, too, is now being affected by acidification of freshwaters. Many of the country's most important fishing rivers are showing signs of declining catches, attributed to acidification (*The Times*, 19 January 1983; Stoner, Gee, and Wade 1984). In the early 1970s, for example, Lake Berwyn (near Tregaron in Dyfed) supported a fine trout fishery; the lake is now fishless. The headwaters of many game fishing rivers are now endangered – including the Conwy, Clwyd, Seiont, Dyfi, Teifi, Tywi, Rheidol, and the upper reaches of the Wye.

In Britain, as doubtless elsewhere, the full picture has yet to emerge because many sensitive areas have yet to be surveyed in detail.

THE HEAVY METAL CONNECTION

The most dramatic symptoms of acidification of surface waters are the massive fish deaths that have been reported from a number of areas. The cause of such deaths is understandably of particularly pressing scientific interest, in light of the media coverage of such events and of the need to minimize as far as possible the risk of further fish deaths.

The early belief was that fish deaths and observed ecological changes in lakes and rivers were caused directly by the acids (including sulphates and nitrates) that had entered the water body. But in the 1970s acidified lakes in Scandinavia and North America were found to have abnormally high concentrations of dissolved aluminium (Al), a toxic heavy metal (a heavy metal being a metal of high specific gravity, normally above five; this means that it is more than five times as dense as water) (Cronan and Schofield 1979; Howells 1984). Leaching of metal ions from soils and rocks in the catchment accounts for the high concentrations of aluminium and of other heavy metals, such as cadmium (Cd), mercury (Hg), iron (Fe), and sometimes zinc (Zn), commonly found in acidified lakes (Dickson 1978). Heavy metals become more soluble in the presence of acids, so they are mobilized by acidification. When water in clay soils reaches pH 4.0, inorganic aluminium is liberated from the soil and is then leached into lakes, streams, and groundwater (Bryce-Smith 1985) (Figure 4.8).

Increasing concentrations of aluminium and manganese (Mn) have accompanied falling pH over a fifteen-year period in some lakes in southern Finland (Wright 1983: 141). Detailed measurements of water chemistry in the Hubbards Brook Experimental Forest in New Hampshire, USA, have shown that acidification causes increased concentrations of aluminium, cadmium, magnesium, and potassium, mobilized from sediments on the stream bed (Likens 1985). Some of the heavy metals are

Figure 4.8   Increase in aluminium concentrations in acidified surface waters in Belgium

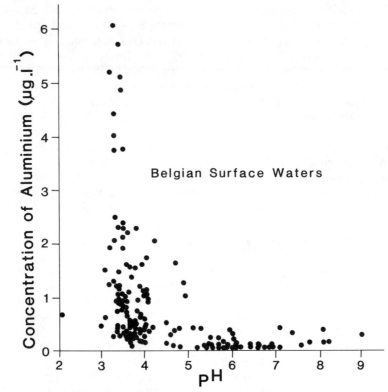

Belgian Surface Waters

Source: after Vangenechten (1983)
Note: Based on data from acid boglakes and less acid fishing pools ($n=151$)

highly toxic (particularly aluminium, cadmium, zinc, and mercury), and they can be passed along the aquatic food chain. Thus heavy metals can be taken into the bodies of aquatic organisms including fish, so causing problems.

Recent studies (Muniz and Leivestad 1980b; Brown 1983) have shown that aluminium is acutely toxic to fish at pH levels that are not normally harmful; aluminium concentrations as low as 0.2 milligrams per litre are known to kill fish (Tolba 1983: 117). Many of the mass mortalities of fish recorded in some Swedish lakes and during spring snow melt have been attributed to aluminium poisoning rather than high acidity alone (Grahn 1980). The physiological response of fish to acid stress is influenced by the toxicity of aluminium. Whilst aluminium poisoning interferes with normal reproduction in fish (Fromm 1980), this effect is now believed to be much less important than the effects of damage to fish gills. Post-mortems on dead fish have revealed low levels of sodium and chloride in the blood, and indicate that the fish died from failure to regulate their body salts because of impaired salt uptake in their gills. Aluminium precipitates on the gills, where it interferes with the normal processes by which oxygen and ions (such as sodium and calcium) are transported across the gill membrane into and out of the fish. This disturbance of ionic regulation within the fish in turn affects the normal transport of gases between respiratory organs and the tissues within the fish (Muniz and Leivestad 1980a; Fromm 1980). Studies of fish deaths in the Adirondack Mountains (Schofield 1976; Cronan and Schofield 1979; La Bastille 1981: 670–1) have established how aluminium damages the fish in two ways – by reducing ion exchange through the gills, and subsequently by salt depletion (through ionic disturbance) and respiratory stress (because of clogging of the gills by mucus) (Muniz and Leivestad 1980b; Driscoll et al. 1980). The fish fry exude mucus to combat the aluminium compounds which collect in their gills, and the mucus collects in large quantities and eventually inhibits the ability of the fish to take in oxygen and salts.

Aluminium can also affect fish in acidified lakes and rivers by reducing their available food supplies. Primary production (the growth of plants) falls if the availability of phosphates in the water decreases (Tolba 1983). Phosphates are often a critical limiting factor in lakes and rivers because they provide essential food for phytoplankton and other aquatic plants. However, phosphates can be complexed (i.e. attached) to the mobilized aluminium. If this happens, a chain reaction is set in motion. Supplies of available phosphates are limited, and so primary production in the water is reduced. This decreases the available food for animals (including fish) higher in the food chain, so there is only enough for a smaller fish population. Decreased plant growth also reduces the number of ecological niches (literally the number of biological 'homes') in the lake or river. Competition for suitable space accompanies competition for available food resources, and further limits animal populations.

One possible means of reducing the fish deaths from aluminium mobilization is to add lime to acidified lakes and rivers. This serves to increase the buffer capacity of the water body, and hence neutralize the acidity. Lime has been added to a great many acidified lakes in Sweden since the mid-1970s, but this has proved to be a costly measure. More preferable, viable, and sustainable are schemes designed to reduce acid inputs before they can start to have an adverse effect on the chemical balance of the surface waters and soils in the catchment (see chapter 7).

CONCLUSIONS

Acidification is now known to affect surface waters in a variety of ways. Acid deposition (wet and dry) alters water chemistry in lakes and rivers, and is associated with declining fish populations in many areas (especially Scandinavia and eastern North America, and more recently parts of Scotland and Wales). The most acute damage has been observed in unbuffered areas, and in association with acid surges (e.g. on spring snow melt or after prolonged drought). The leaching of acid minerals from soils into surface waters compounds the problem.

Some freshwater species are very sensitive to large inputs of acid, but the primary cause of death to most fish is now known to be poisoning by toxic aluminium, which is mobilized in soils by acid fallout and then washed from the soils. Acidification triggers biological changes at all levels in the aquatic food web. Acidified lakes and rivers have smaller and less diverse populations of most species; microscopic plants and animals decline, most large plant life is seriously affected, most large fish decline. *Sphagnum* moss and filamentous algae increase, and form impenetrable mats on the lake beds.

A growing number of scientists of different nationalities conclude that acid deposition is having serious impacts on fish populations and on aquatic ecosystems, and a consensus is emerging that the primary culprit is probably the emissions of sulphur and nitrogen oxides from power stations, factories, houses, and vehicles. However, whilst the links between oxide emissions, downwind transfer/chemical transformation/deposition, acid deposition, and damage to freshwater ecology appear to be established with at least some scientific objectivity, it remains impossible to obtain unequivocal evidence that is convincing to the most entrenched (but generally scientifically unenlightened) politician.

However, such is the nature of scientific enquiry and the complexity of the natural world that conclusions must always be drawn on the basis of probability – e.g. the 'great likelihood' that (a) causes (b), and that more times than not a change in (a) will trigger a corresponding change in (b). In the case of acid rain damage to surface waters, the balance of probability currently rests firmly in the court of those who argue that reductions in oxide emissions (especially from power stations) will bring reductions in the spread of damage to lakes and rivers, although they argue that such a reduction alone cannot help to repair the damage already inflicted on many lakes and rivers in Scandinavia, North America, and Great Britain.

5 EFFECTS ON SOILS AND VEGETATION

'Deep in the forest . . .'

The thirsty earth soaks up the rain,
And drinks, and gapes for drink again.
The plants suck in the earth, and are
With constant drinking fresh and fair.

(Abraham Cowley, *Drinking*)

Forest damage and fish deaths have battled for media attention as the most serious problem widely attributed to acid rain. The forest issue first hit the international headlines during the early 1970s, when damage to both coniferous and deciduous trees was observed in West Germany for which there was initially no obvious explanation (La Bastille 1981: 674). The possible extent of the problem became clearer during the early 1980s, when German foresters were able to document a dramatic increase in the pace of tree dieback and reports started to appear of similar damage in Scandinavia and parts of Britain.

This chapter will examine the evidence of damage to trees and crops, and evaluate the suggested links with acid rain. Acidification also alters soils (by changing soil chemistry and mobilizing toxic heavy metals), and so the links between acid rain, soils, and damage to trees and crops must also be explored.

The ecological effects of acid rain on lakes and rivers are easier to establish than those on soils, trees, and plants. This is partly because of the wide range and large number of possible interactions between the atmosphere and terrestrial ecosystems, and partly because effects on soils and trees may take some time to reach detectable dimensions (perhaps decades in the case of trees). Effects on crops have important economic and agricultural implications, but those on trees have attracted more attention to date (hence they are our main concern in this chapter). Paradoxically, effects on trees are more difficult to detect than those on crops because of the greater biological complexity of forest ecosystems than of croplands, and also because forest systems have much longer response times to acid stress.

ACIDIFICATION OF SOILS

Concern over the possible acidification of soils has risen in recent years because soils support the plants and animals that give us food, fibre, and timber products. Fears are growing that, if air pollution continues unchecked, vast areas of sensitive soils may slowly decline in fertility until the critical point at which productivity fails altogether – after which biological damage may be extremely difficult to reverse, at any cost (La Bastille 1981: 674).

Soils are generally much less vulnerable than surface waters to acidification. This better resistance reflects the buffer capacity of the soils, and means that most soils can tolerate higher levels of acidity than lakes and rivers, without visible damage. Vulnerability varies, however, depending on soil type, underlying bedrock, and human use and management of the soil (Malmer 1976; Tolba 1983: 117). Soils on bedrock that has a low calcium content (e.g. granite) and thin soils with low buffering capacities (see chapter 3) are the most vulnerable to acidification. In such soils acidity can build up to biologically intolerable levels. In contrast, in alkaline soils (e.g. over limestone), acids can be absorbed then leached out in the form of salts by water draining through the soil (Buckley-Golder 1985). Soil management also plays a key role. The less acid (pH > 4.5) soils are often subject to intense agricultural management, including fertilizer applications, drainage, and liming. Deterioration of soil quality in many agricultural areas has been prevented by the application of lime (a calcium product), which will normally also serve to neutralize the effects of pollutant acidity (Last and Nicholson 1982: 250).

Complexities and complications

The effects of acid deposition (wet and/or dry) on soils are often difficult to detect, for various reasons. One is that acids can also be created by some natural processes within the soil (Last 1982). These include biological processes (e.g. the breakdown of dead organic matter) and the weathering of mineral soil particles. Some soil scientists now argue that the addition of acids to many soils through acid deposition is almost insignificant compared with the levels of production of acids by natural soil processes and agricultural soil management (Krug and Frink 1983; Tabatari 1985). As a result, they argue, the possible interactions between acid rain, acid soil, land use, and vegetation on a watershed need to be carefully assessed when evaluating the benefits that can be expected from proposed reductions in SO_2 and NO_x emissions (see Part IV). A second complication is that acid deposition (at low levels of acidity, and in nutrient-deficient soils) can produce apparently beneficial effects on soils and vegetation, acting as a fertilizer via the provision of free nutrients (especially of sulphur and nitrogen) (Abrahamsen, Stuanes, and Tveite 1983).

In many areas, acidification (whether from natural processes or via acid deposition) damages soils by altering their microbiology, chemistry, and fauna. There are two main biological effects of increased soil acidity. One is a reduction in the rate of bacterial decomposition, through a slowdown in the organisms that break down litter on the soil surface. The second is a reduction if not limitation in the rate at which plant roots can absorb soluble material (such as soil moisture and dissolved nutrients). Both effects can produce significant changes in soil structure and productivity, and in turn in vegetation.

A complex series of chemical reactions takes place when soil is acidified, the most important of which are believed to be the replacement of basic cations (K^+, Ca^{++}, Mg^{++}, Na^+) by hydrogen and aluminium ions. This cation exchange process produces two chemical effects that further alter the soil and in turn produce problems for the plants growing in that soil. The primary effect is to increase the leaching of cations from the soil and thus reduce the availability of nutrients within the soil. The secondary effect is to dissolve and mobilize (thus liberate) toxic heavy metals that are present but normally chemically 'bound up' (thus not freely available) in the soil. Acidification may also increase the weathering of silicate minerals in some soils, leading to loss of mineral structure and possible soil impoverishment (and reduced fertility).

Nutrient deficiency

One immediate impact of an input of acids is to increase the exchange between hydrogen ions and the nutrient cations potassium (K), magnesium (Mg), and calcium (Ca) in the soil. As a result these cations are liberated within the soil, and they can be rapidly leached out in solution in the soil water along with the sulphate from the acid input (Abrahamsen, Stuanes, and Tveite 1983; Tolba 1983; Van Breeman, Driscoll, and Mulder 1984). An increase in soil acidity (decrease in soil pH) normally brings increased leaching of potassium, magnesium, and calcium, and the greater the increase in acidity the greater the loss of these nutrients from the soil.

Poorly buffered sites often show a progressive loss of base cations (especially exchangeable Ca and Mg) through time. For example, there are records of decreasing amounts of potassium, magnesium, and calcium in the soils beneath some Swedish forests over a ten-year period during which acid deposition promoted a decline in soil pH (Abrahamsen, Stuanes, and Tveite 1983).

Plants require these macro-nutrients for growth and survival, and they take them in dissolved in soil moisture through their roots (by root osmosis), in variable but generally relatively large quantities. Consequently, acid-induced leaching leads to nutrient deficiency in the affected soils, and this loss of soil fertility leads to a slowdown in the rate of growth of plants and trees on acidified soils.

Nutrient availability within trees (and other plants) is also influenced by chemical exchange processes that take place on the leaf surface (Abrahamsen *et al.* 1976). Ammonia and nitrogen, which land on the leaf via acid deposition (wet and dry), pass through the semi-permeable membrane on the leaf surface and are incorporated within the leaf cells. There a chemical interaction (cation exchange) takes place, and those nutrients that are available in abundance (e.g. potassium, calcium, magnesium, and sulphur) are leached from and get washed off the leaf surface. It is *not* clear whether this foliar leaching process significantly affects the nutrient status of the trees or not. It is believed that the adverse effects of reduced nutrient uptake via soil moisture far outweigh the problems of foliar leaching.

Natural undisturbed ecosystems have complex nutrient cycling mechanisms that ensure that essential nutrients are available in sufficient quantities and in suitable forms. Thus the nutrients stored in living plants are released back to the soil when the plant dies and its body is decomposed by bacteria and micro-organisms within the soil. If this cycle of growth and decay is allowed to proceed uninterrupted, natural biological mechanisms ensure long-term stability in nutrient availability. But when part or all of the living products in an ecosystem are cropped (e.g. timber harvesting in forestry, or the harvesting of agricultural crops), the nutrient cycle is short-circuited and there is a net loss of nutrients (in the harvested produce). Fears have been expressed over the real prospect of nutrient depletion where traditional forest felling practices are replaced (as in parts of eastern Canada) by whole-tree harvesting techniques. In traditional practices only the main tree trunks are removed, and all other dead wood is burned or left to decay on site (thus replenishing at least part of the nutrient loss). In whole-tree harvesting techniques, in contrast, all material is removed from site and there is a wholesale loss of nutrients. Over the long term such accelerated nutrient depletion may lead to declining forest yield and further increase the susceptibility of tree growth to the impact of acid deposition.

Acidification can also affect nutrient cycling (thus availability of nutrients) in soils by changing the rate at which dead organic matter is broken down and consequently the rate at which nutrients are liberated back to the soil. Experimental studies show that strong acidification retards the decomposition of litter of spruce, pine, birch, and other cellulose-rich materials (Tamm, Wiklander, and Popovic 1977; Francis 1982; Killham, Firestone, and McGoll 1983). This reduction in rates of decomposition can in turn produce several effects, the most important of which are:

- reduced respiration by organisms within the soil (including bacteria),
- increased levels of ammonia in the soil caused by a reduction in the mobilization of nutrients (previously released in decomposition), and
- decreasing soil nitrate levels (because of ammonification).

Such changes in soil chemistry may exceed the tolerance limits (see chapter 4) of many soil animals, and so population changes may be expected in acidified soils. Experimental studies (Abrahamsen, Stuanes, and Tveite 1983) with soils at pH below 3.0 show a decreasing total abundance of enchytraeids, likely increases in the abundance of springtails, and no observed changes in the abundance of mites.

Mobilization of heavy metals

Increased soil acidity is often closely associated with increased concentrations of some toxic heavy metals. The most common are aluminium (Al), cadmium (Cd), zinc (Zn), lead (Pb), mercury (Hg), manganese (Mn), and iron (Fe) (Tolba 1983: 117). These metals, derived from natural weathering processes and stored within the soils, are mobilized (i.e. they become more soluble in soil water) in the presence of acids. This mobilization encourages the rapid spread of the metals through soils, in natural flows of soil water. From soils they can be washed into lakes and rivers (see chapter 4) and taken in by plants and trees (via root osmosis). The heavy metals can also contaminate groundwater and be washed into reservoirs, and in turn passed on to humans (see chapter 6).

However, as with nutrients, the position can produce either benefits or problems, depending on the concentrations involved and critical thresholds of biological activity. Some metals (such as Mn and Fe) provide essential micro-nutrients that stimulate plant growth at low concentrations. However, at higher concentrations the same metals can be extremely toxic and greatly reduce plant growth.

There is now little doubt that an increase in soil acidity brings higher amounts of aluminium and other heavy metals in soils. But the critical question remains: 'What damage does it cause?' Scientists are divided over the issue, and much basic research remains to be done before categorical conclusions can be drawn. It is known, for example, that many tree species have adapted to an acidic soil environment with high aluminium concentrations (Abrahamsen, Stuanes, and Tveite 1983), and they can apparently tolerate quite high concentrations of some heavy metals in the soil. However, one view, based on findings of large amounts of aluminium in soils in damaged forests of central Europe (Ulrich, Mayer, and Khanna 1980), is that increased concentrations of aluminium damage plants directly. The Ulrich theory is that aluminium damages the fine root hairs of plants and this reduces the uptake of phosphorus and other essential nutrients. In turn the plant withers and may eventually die – if not from nutrient starvation then possibly from reduced resistance to severe weather (cold or drought) or disease.

It is difficult to speculate about whether or not such damage might emerge in Scandinavian forests, because as yet relatively little is known in detail about aluminium concentrations in forest soils there (Abrahamsen, Stuanes, and Tveite 1983), or about the heavy metal tolerance of the Norway spruce (one of Scandinavia's most important tree species). But the dramatic loss of fish in many lakes and rivers in southern Norway and Sweden (chapter 4) does indicate a clear likelihood of heavy metal mobilization and inwash into surface waters.

Similarly, recently observed damage to lakes and rivers in parts of upland Britain (particularly in Wales and Scotland) suggests that unacceptably high concentrations of heavy metals (especially aluminium) might well be involved.

EFFECTS ON VEGETATION

It has long been known that high concentrations of sulphur in the atmosphere can damage vegetation. For example, in 1913 experiments were carried out on the growth of lettuces in different parts of Leeds in north-west England, which showed a close connection between variations in the amount of sulphate in the air over different parts of the city and variations in the size and quality of lettuce grown (Goudie 1984). More recently, vegetation 'deserts' and clear spatial zonations of vegetation around point sources of SO_2 pollution (such as aluminium smelters in Norway – Gilbert 1975) have been documented in many areas. Tree canopies are often reduced if not totally absent within about 6 km of smelter plants, and few pine trees are found within 25 km of such point sources (Woodwell 1970).

Most of the early interest centred on such local biological effects. Since the mid-1970s, however, interest has broadened to include the possible effects on vegetation of long-range (perhaps trans-frontier) deposition (wet and dry) of acid materials.

Lichens as air pollution indicators

The best biological indicators of air quality are lichens – flat plants that occur on a wide range of types of surface (including tree trunks, rocks, walls, and the ground). Devoid of roots, lichens rely for nutrients on dissolved solids in rainwater that washes over them. The water and nutrients are absorbed through the whole body surface of the lichen (Mabey 1974: 102–5); the lack of buffer capacity (in the absence of soils) means that lichens are directly exposed to potentially high levels of acidity in rainfall.

Lichens (especially the corticolous species) are very sensitive to (thus intolerant of) SO_2 poisoning, and they provide a sensitive 'biological litmus' indicator (Ferry, Baddeley, and Hawksworth 1973): the worse the SO_2 pollution, the fewer the types of lichen that can survive. It is not known exactly how acidity affects lichen communities, but it is believed that two mechanisms are involved (Hale 1974; Arvidsson 1985; Berlekom 1985d). Minerals in the substrate material exist in different chemical states under different chemical regimes, so that nutrient availability to the lichens will probably vary with varying rainfall acidity. Moreover, it is likely that the rate of nutrient assimilation in lichens varies with changing pH.

Lichens have rarely been found in the centres of major cities, such as Bonn, Helsinki, Stockholm, Paris, or London (Goudie 1984), which are

– or have been in the recent past – often exposed to high concentrations of SO_2 from traffic, industry, power-generating utilities, and domestic chimneys. Improvements in air quality that have followed the introduction of air pollution legislation have been accompanied in many cities (e.g. London since 1960 – see chapter 1) by reinvasion of lichen species.

Lichen distributions also provide a means of examining air pollution over wide areas. Field studies indicate that by 1970 epiphytic lichens (which grow on other plants, mainly trees) had disappeared from about a third of England and Wales (in a belt running from London to Birmingham, through the Midlands, and broadening to include most of Lancashire and West Yorkshire and parts of Tyneside – the traditional heavy industrial axis) (Hawksworth and Rose 1970). Recent studies of lichen populations in the Borrowdale woodlands in the central part of the English Lake District (Taylor 1985) indicate an increasing abundance of acid-tolerant lichen species accompanied by a reduction in the presence of acid-sensitive lichens, attributed to acid deposition.

Trees and higher plants

Higher plants (herbs and grasses, shrubs and trees) are also adversely affected by SO_2 and acid deposition. The damage arises in two ways – through foliage (which is directly exposed to acid rain and fogs, and to dry fallout of oxides and particulates) and through roots (which are affected by soil nutrient deficiency and aluminium poisoning).

Visible damage to vegetation can assume various forms depending on the character and level of the acidification, and the buffer capacity of the area. Common symptoms include direct physical damage to plant tissues (especially roots and foliage) – reduced canopy cover, crown dieback, and whole tree death (Tamm and Cowling 1977; Tomlinson 1983; Buckley-Golder 1985) – and reduced growth rate of plants (hence declining crop yields in farming and forestry). A typical sequence of symptoms of progressive dieback of British beech trees is shown in Figure 5.1.

However, symptoms vary between species, and even within a given species it is often very difficult to recognize the symptoms of damage. Injury to vegetation is not always manifest by visible damage to leaves; it can occur as reduced plant growth, reduced capacity for flowering, and

Figure 5.1   Typical stages in the dieback of British beech trees

| Stage 1 (normal) | Stage 2 (partial dieback) | Stage 3 (advanced dieback) | Stage 4 (complete dieback) |

Source: after Anon (1985)

reduced yield, and each of these symptoms might take some years to spot with certainty. Early signs of damage can be overlooked even by the keen and well-trained eye, so that adverse effects only become apparent when they reach critical and perhaps irreversible proportions.

Germination and growth rates

Natural regeneration of plants is one of nature's inbuilt self-repair mechanisms, and it is of critical importance if damaged plants are to be replaced. Experimental evidence is available for Norway spruce, Scots pine, and silver birch seeds (Abrahamsen, Stuanes, and Tveite 1983) that indicates that germination rates are moderately inhibited when soil pH is between 3.8 and 5.4. The germination of birch seeds is inhibited at lower soil pH.

The initial establishment of seedlings appears to be more sensitive than germination to soil pH, and it is known that the establishment of Norway spruce, Scots pine and silver birch decreases rapidly when soil pH drops below 4.2. For this reason, reforestation of areas damaged by acid deposition is often not possible, because new trees struggle for a few years then eventually they lose their vitality in the poisoned air and soil (Halford 1986: 16).

Growth rates are best measured by the annual rate of incremental increase of tree rings. There is evidence from some forests of declining widths of annual growth rings through time, even in healthy trees (see Figure 5.2). But studies of tree-ring growth rates in sample areas of Norway and Sweden with different levels of acid input found no unambiguous trends, and growth rates were not found to be reduced systematically in areas with high rates of acid deposition (Abrahamsen *et al.* 1976). However, neither is there 'sufficient evidence to state that acid precipitation has no effects, as further acid deposition on forest ecosystems

Figure 5.2   Declining annual growth-ring widths in damaged silver fir trees in the Alpirsbach forest district of West Germany, 1940–83

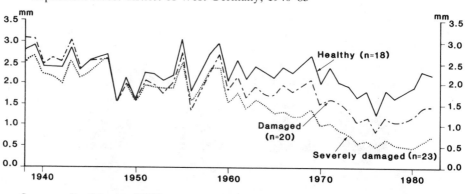

Source: after Schroter (1985)

could cause unknown changes in the chemical and biological properties of the soil that could affect its overall productivity' (Lesinski 1983: 158).

A common pattern in tree growth studies is to witness increased growth of tree height and trunk diameter in the early years of experimental plots (with acidity levels similar to those presently found in Scandinavia) compared with similar plots with no acids added, followed by either no difference between the two sets of plots or a reduced rate of diameter growth in the acidified ones (Abrahamsen, Stuanes, and Tveite 1983). The initial growth increase probably reflects the stimulatory effect of nutrient provision (see Figure 5.3).

Physical damage to trees

It is the visible damage to trees that has given rise to most concern, especially in West Germany, parts of the Eastern United States, and Canada, and more recently in Scotland. Visible damage tends to be concentrated in older and more established trees, and it appears to be species specific. One of the most sensitive tree species is Scots pine. This is a very significant component of the forest economy in many countries (such as Poland – Lesinski 1983: 158), so that acid-induced forest damage may have serious consequences for national economies and some aspects of international trade in raw materials.

The exact form and severity of the damage varies according to the character and magnitude of the air pollution, so that visible evidence of

Figure 5.3   Initial growth increase followed by slower growth rates in trees under acid deposition

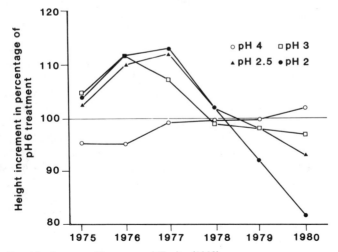

Source: after Abrahamsen, Stuanes, and Tveite (1983)
Note: Based on field experiments in a Scots pine sapling stand, in which trees were watered with 'acidified' rain

physical damage varies considerably from place to place. Symptoms of damage appear with varying intensity, and in various combinations. The most common symptoms of disease shown by damaged conifers (see Figure 5.4) are yellowing needles, loss of needles, distorted branches, thinning tops, bark injuries, trunk changes, and damage to fine root fibres (Agren 1984a: 6). Common tell-tale signs in Scots pine (Lesinski 1983) include:

- needles become shorter
- durability of the needles decreases (from three years to one year)
- top buds dry up
- annual increments of shoot length become smaller
- the shape of the tree crown changes.

In deciduous trees (see Figure 5.1) the main visible symptoms are discoloured and deformed leaves, early shedding of leaves, death of part of the tree top, bark injuries, and a lack of natural regeneration (Agren 1984a: 6).

Under extreme conditions, the tops of injured trees (deciduous and coniferous) die earlier than the branches further down, giving a curiously dismal appearance to some forests. This form of crown dieback has been particularly pronounced in West German forests.

WALDSTERBEN – DAMAGE TO WEST GERMAN FORESTS

World attention has been focused on the forests of West Germany, where tree damage has been most acute and is now well documented, and which appear to have suffered from dramatic increases in the level and type of damage in recent years.

Tree damage first came to light when diseased fir trees were found in the Black Forest in the early 1960s. What was initially viewed as probably a local problem had clearly intensified by the late 1970s, when some experts concluded that one-third of the fir trees had died already and half of the remaining ones may be terminally ill (Halford 1986: 15), and fears were expressed for the future of the Black Forest (May 1982: 11).

Coniferous trees (fir and spruce) in the Black Forest have been worst affected, and 'instead of lush, dark green foliage that gave the Schwarzwald [Black Forest] its name, many trees now have only sickly yellowy-brown needles' (Halford 1986: 15). But other tree species in the forest are also showing signs of damage: half the beech and 43 per cent of the oaks are affected, and many elm trees are dying at 60 years old rather than the more normal 130 or so years (Halford 1986: 16).

The damage appeared to be caused by some form of disease described as a 'baffling form of tree "cancer" ' (Tifft 1985: 45), and it produced striking visible symptoms in affected trees. Damage in the spruce, for example, was vividly described by *Time* magazine:

the dark green branches at first hang limply, like Spanish moss. Between three weeks and five years later, the branches are tinged with

Figure 5.4 Typical symptoms of damage in coniferous trees subject to acid rain, with assumed processes

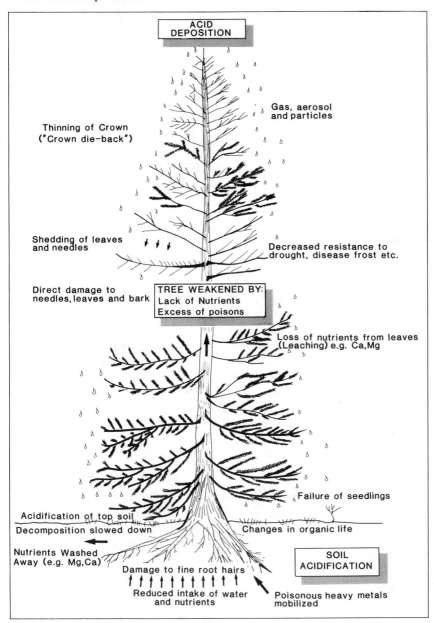

Sources: modified after Binns *et al.* (1985), Dudley, Barrett, and Baldock (1985), Swedish Ministry of Agriculture (1984), and other sources

yellow and then brown. The weakened tree soon drops its needles and eventually stops growing new ones. It becomes bald at the top, and appears stunted, spreading its branches outward and upward like a stork's nest. In a desperate struggle for life, it may grow excessive numbers of cones or sprout 'anxiety shoots' – tiny branches that grow irregularly along the bough. Roots and trunks begin to twist and shrink. Finally, drought, frost, insects and parasites finish off the weakened plant. In the end, it stands like a bony finger pointing toward the sky.

(Tifft 1985: 46)

It was during the autumn of 1980 that German scientists first observed the mysterious wasting disease – the so-called *Waldsterben* (meaning literally 'death of the trees' or 'dying forest syndrome') – that blighted trees and forests. Professor Ulrich, the biochemist who first reported extensive tree damage in West Germany, attributed it to acid deposition (Ulrich, Mayer, and Khanna 1980) and argued that unless steps were taken to reduce acid deposition damage would escalate and become irreversible. Ulrich identified three stages in the damage process (see Figure 5.4) (Pearce 1982a):

(1) Nitrates or nitrogen oxide in the acid rain provide nutrients to the soil, thus trees grow more rapidly. The acid deposition ends up in lakes and rivers, leading to fish deaths.
(2) Soils progressively lose their ability to neutralize the increased acidity, and acids begin to accumulate and combine with other nutrients leached out of the soil. This slows down tree growth and may lead to yellowing of pine needles. Sulphate combines with metals in the soil (especially aluminium) and mobilizes them into toxic solutions.
(3) Toxic aluminium is released (at pH 4.2), which leads to destruction of tree roots and deterioration of natural defence mechanisms within the trees that prevent the entry of bacteria, fungi, and other viruses. Slow death of the tree through starvation, disease, and toxic poisoning ensues.

German forests were faced with a series of problems, including the mystery wasting disease, increased storm damage, and regeneration difficulties – all attributed either directly or indirectly to acid deposition. West German forests are believed to receive more acid fallout than Scandinavia, being located close to many large cities and industrial centres (such as the Ruhr corridor) with abundant emission sources.

The speed at which damage appeared to be spreading was alarming. In 1982 only 7.7 per cent of West Germany's 7.4 million hectares of forest were visibly damaged; within a year 34 per cent of trees had suffered discolouration and some loss of needles and leaves; by late 1984 around half the country's woodlands showed symptoms of the disease (Tifft 1985: 47). Table 5.1 shows that the greatest absolute damage had been

Table 5.1   *Spread of forest damage in Federal Republic of Germany, 1982–5*

(a)   *Forest damage by tree species*

Species	% of forest area	% of trees damaged			
		1982	1983	1984	1985
Spruce	40	9	41	51	52
Pine	20	5	44	59	58
Silver fir	2	60	75	87	87
Beech	17	4	26	50	55
Oak	8	4	15	43	55
Others	13	4	17	31	31
Total	100	8	34	50	52

(b)   *Forest damage by category*

Category	% of forest area			
	1982	1983	1984	1985
1. slightly damaged	6	25	33	33
2. damaged	1.5	9	16	17
3. severely damaged/dead	0.5	1	1.5	2.2
Total	8	34	50	52

Sources: Agren (1984a, b), Anon (1986a)

suffered by spruce (over half of the country's 1.2 million hectares damaged by 1984), and the greatest relative damage had been suffered by silver fir (over 87 per cent of the trees were damaged by 1984) (Agren 1984a: 6). Spruce is the most important tree from an economic point of view, and silver fir is 'now regarded as threatened with extinction' (Agren 1984a) in West Germany. The rapid spread of damage through previously healthy trees in Baden–Württemberg in West Germany is clear from Figure 5.5.

During the early 1980s it was believed that *Waldsterben* took between two and three years to kill trees, but by 1985 cases were being reported in which the effects became clear within the bewilderingly short space of five weeks (Tifft 1985: 47). Alarm was also expressed at the apparently unselective nature of *Waldsterben*, which was affecting young saplings as well as mature trees (firs over 120 years). Moreover, individual trees within forests were being affected: 'the dead trees are usually found isolated in healthier stands. Groups of dead trees, or dead forest lots, are very seldom found in West Germany' (Agren 1984b).

The disease was also affecting a growing number of species (Tifft 1985).

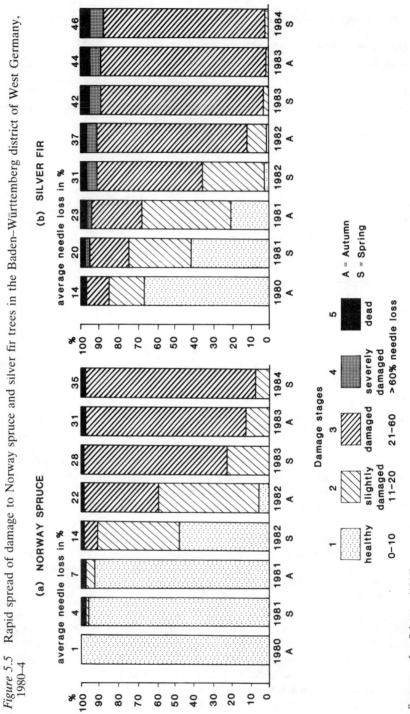

Figure 5.5 Rapid spread of damage to Norway spruce and silver fir trees in the Baden–Württemberg district of West Germany, 1980–4

Source: after Schroter (1985)
Note: Based on data for 556 Norway spruce in 17 observation plots, and 1,675 silver fir in 27 plots

The first signs of disease were discovered in silver fir (*Abies alba*) in the early 1970s, but by the late 1970s damage was also reported in the Norway spruce (*Picea abies*). By the early 1980s damage was evident in pine (*Pinus silvestris*), beech (*Fagus sylvatica*), and to a more limited extent in other species such as larch, oak, red oak, maple, ash, and rowan (Agren 1984a: 6). The rapid spread of damage through species such as oak, beech, and pine during the early 1980s is evident in Table 5.1.

Apart from the amenity (tourism and recreation) and conservation loss, the large-scale and widespread damage to trees has potentially significant financial consequences. Unofficial estimates of the cost of tree death, reduced growth, and lower-quality timber in West Germany are in the region DM 7–10 billion per year (Agren 1984a: 7).

DAMAGE SPREADS

By the early 1980s *Waldsterben* appeared to be advancing across Europe, blighting woods in countries including France, Switzerland, Sweden, and Italy. By 1985 over 30,000 hectares of woodland in France were showing signs of damage, most seriously in the Vosges and Jura Mountains.

There are two possible explanations for this apparent spread of tree damage across Europe: that damage was actually on the increase, or that observers were becoming more aware of what symptoms to look for. In reality it is likely that both trends occurred. The 'better awareness' argument certainly holds true for Scandinavia – Swedish foresters only observed damage effects in 1983 after they had been briefed by the German forest authority (Anon 1984a).

Eastern Europe has not escaped the ravages of *Waldsterben*. An estimated 86 per cent of East Germany's 3 million hectares of woodland and 20 per cent of Czechoslovakia's 960,000 hectares of woodland are showing visible signs of damage (Tifft 1985: 46), and damage is believed to be evident in Poland, Hungary, and the Soviet Union (although reported evidence is as yet scanty).

Damage has not been confined to Europe, however, and there is mounting evidence of sick and dying trees in eastern North America. Some of the best United States evidence is available for Camel's Hump, a peak forested with red spruce and balsam firs in the Green Mountains of Vermont (Vogelman 1982). Research on the changing vegetation of the forest, carried out at the University of Vermont, shows that between 1965 and 1981 half of the spruce trees had died and seedling production, tree density, and basal area (i.e. tree growth rates) had fallen by nearly a half. Wood volume on Camel's Hump fell considerably after 1965, and it was concluded that 'if such losses in only a few years are representative of a general decline in forest productivity, the economic consequences for the lumber industry will be staggering' (Vogelman 1982).

Canada, with a $20 billion forest industry, is understandably very concerned over the potential effects of acid rain (see chapter 9). The discovery in 1984 of sugar maple trees in Quebec Province showing symptoms of severe drought (even in the midst of abundant moisture) (Pawlick 1985), which some experts regard as similar to the evidence of tree deaths in Europe attributed to acid rain, has done little to allay Canadian fears.

FOREST DAMAGE IN BRITAIN

Given the apparent spread of tree damage through Europe during the early 1980s, fears began to mount over the prospect of acid rain damage to trees in Britain. The official view in Britain – expressed by the Forestry Commission, and based on their own field survey carried out in 1984 (Binns *et al.* 1985) – was that there was no evidence of acid rain induced tree damage in Britain. Tests were carried out on Scots pine, Norway spruce, and Sitka spruce, and most were given a clean bill of health.

Dr Bill Binns, leader of the Forestry Commission survey, concluded that 'British conifer forests reveal no symptoms similar to those seen in the damaged areas of western Germany – or rather, any similar symptoms have been around for many, many years, and there is no damage that can be described as "new" ' (W.O. Binns reported in *The Times*, 6 August 1984; Binns 1984a). This statement was very unconvincing to many of Britain's conservationists, given that in December 1984 William Waldegrave (Parliamentary Under-Secretary of State at the DoE) acknowledged the existence of new and unexplained damage to British trees (*Hansard*, Written Answers, 20 December 1984), and in mid-1984 the government's own Select Committee inquiry into acid rain stressed the need to confirm recent reports of damage to British trees that seemed to be similar to that observed in West Germany (*The Times*, 12 June 1984). Moreover, ecologists had been aware for some time that coniferous trees in parts of the Pennines, especially around Manchester (Ferguson and Lee 1983; Lines 1984; Bryce-Smith 1985), were showing signs of damage that might be related to air pollution from sulphur dioxide, and recent attempts to plant conifers there had not always been successful. Evidence was also emerging from woodlands in Cambridgeshire and Essex and forests in the Lake District and Galloway of apparent damage from acidification (Pearce 1982b).

Unconvinced by the Forestry Commission's conclusion, and incensed by their 'lethargic response', Friends of the Earth (Scotland) carried out Britain's first survey of acid rain damage to trees during 1984 (Anon 1984a). It was a partial survey, based on forty-six sites in England and one in southern Scotland, and led by Joachim Puhe of the University of Göttingen, West Germany. The results were startling: thirty-one of the sites showed signs of damage to conifers and hardwoods, and a range of species were affected (Douglas fir, Norway and Sitka spruce, beech, and pine). The most acute damage was observed in the Lake District (advanced needle loss on Sitka spruce around Whinlatter Pass) and in the Forest of

Dean near Avonmouth (canopy loss, branch deformation, and loss of leaves on beech). FoE (Scotland) immediately called for an urgent and comprehensive tree survey to evaluate the extent of damage and potential loss of forest revenue. This was to be the start of a prolonged series of exchanges of rhetoric between FoE and the Forestry Commission, both of whom challenge each other's results.

Conscious of the need to study British tree damage in much greater detail, and to have expert guidance and at the same time to maximize possible media exposure, Friends of the Earth (with World Wildlife Fund sponsorship) flew in Swedish forest ecologist Dr Bengt Nihlgard in early August 1985 (*The Times*, 15 August 1985). Nihlgard found signs of damage to seventeen British tree species (especially deciduous ones) from eight counties, comparable to, if not worse than, what he had seen in Sweden and northern and central Europe. Yew trees were found with characteristic yellowing and early needle loss (the so-called 'tinsel syndrome'); beech showed typical reduced crown growth, pale yellow leaves, and early leaf drop; some Norway spruce had high (over 50 per cent) needle loss, comparable to that seen in the worst-hit parts of Europe (Anon 1985).

On 15 August 1985 the first national survey of acid rain damage to native British trees was launched by Friends of the Earth. The national survey, also funded by the UK World Wildlife Fund, was inspired by the findings of the earlier partial survey and the results of Nihlgard's visit. The brief – to examine damage to yew and beech trees that might stem from acid rain – was prepared because environmental groups believed that official denial of damage to young forestry plantations overlooked widespread dieback in older native trees and hedgerows (Anon 1985). The 1985 FoE survey was a direct challenge to the official British opinion that native trees were not being adversely affected by acid rain. The Forestry Commission were quick to seek to undermine the basis of the FoE claim, pointing out that their own (1984) survey of deciduous trees in the New Forest found no evidence of air pollution damage. Binns stressed that 'the damage done to trees by the droughts of 1976, 1983 and 1984 still shows. That may be the reason for the confusion' (*The Times*, 23 August 1985).

By late August 1985 there was clear evidence of tree damage to deciduous (especially beech) trees in Britain. Neither party doubted that the evidence existed, but interpretations differed. The Forestry Commission blamed climatic factors (especially the legacy of the 1976 drought) and FoE blamed acid rain. Anxious to avoid a 'war of words' with Friends of the Earth, the Forestry Commission agreed to carry out a more detailed sample survey of beech trees (*The Times*, 26 August 1985). Tentative results of the survey were released in September 1985, and suggested no significant link with acid rain. The Commission insisted that observed damage could be attributed to drought, fungi, or insects (Forestry Commission 1985; Rose 1985).

The results of the FoE *Tree Dieback Survey* were published on 6 October 1985, and were interpreted as clear signs of acid rain damage to beech and yew trees in England and Wales. Only one-third of the 3,184 beech and yew trees examined at 799 sites in 49 counties were considered to be healthy, ten other species (including fir, oak, spruce, and pine) were found to be at risk, and damage similar to that blamed on acid rain in Europe was found to be widespread. The survey described in graphic detail the 'death of an ancient yew', in four stages (*The Times*, 1 November 1985):

(1) The tree is in good health, with green needles.
(2) Patches of thinning appear on the tree, older needles (further up the twig) often turn yellow then drop off. The tree is covered with ragged, drooping twigs (the so-called 'tinsel syndrome').
(3) Main branches become visible within the tree as needle loss intensifies, tinsel syndrome causes a 'net curtain' effect, the canopy is semi-transparent and casts little shade (plants can now establish themselves beneath the tree).
(4) The tree is moribund, branches have a mossy appearance, the tree silhouette is skeletal.

Friends of the Earth claimed, proudly and defiantly, that 'we think it shows that acid rain damage has definitely arrived in Britain' (Chris Rose, reported in *The Times*, 7 October 1985). Professional foresters in Britain – insistent that observed effects were compatible with drought damage – were not convinced, and accused FoE of publishing a 'scaremongering' report, based on 'amateurish founderings', which 'achieved nothing but publicity' (correspondence from Dr B. N. Howell to *The Times*, 17 October 1985).

The tree damage debate in Britain has quietened down somewhat since late 1985, but many ecologists and environmental groups are monitoring woods and forests in search for more conclusive evidence of the effects of acid deposition on both deciduous and coniferous trees. In the final analysis, the two sides in the debate are not necessarily forwarding mutually exclusive or diametrically opposed interpretations, because it is quite possible (if not likely) that British tree damage is being caused by climatic effects on trees with reduced resistance as a consequence of acid damage.

COMPLEXITY OF FOREST DAMAGE

There is now little doubt that a growing number of individual trees and even whole forests are dying, especially in Europe. Various explanations have been offered for such widespread damage, and the most convincing evidence points to air pollution – and especially to acid deposition – as a prime catalyst. The consensus view is now emerging amongst many European and North American scientists that, even if air pollution is not the only factor involved, tree damage and death are unlikely on such a large scale without air pollution. Trees are attacked by acid deposition on many fronts (see Figure 5.4), and 'when death comes, it is often due to natural causes – insects, high winds, frost and drought. Much like an AIDS victim whose immune system has broken down, the ailing tree is left defenseless against the ravages of nature' (Tifft 1985: 47).

The evidence suggests that the damage is caused by acid deposition from air pollution in combination with a whole host of contributory factors (Lesinski 1983: 158; Tolba 1983: 117; Ribault 1984; Blank 1985). These include:

- the dry fallout of SO_2 and NO_x
- ozone
- heavy metals
- parasites and plant diseases
- extreme climatic conditions (very high and very low rainfall, temperature extremes – especially frost)
- forest management practices (e.g. forest fire prevention in some areas has kept alkaline ash from neutralizing some acid-polluted soils)
- site factors (such as soil drainage)
- the nature of underlying soil
- the general state of health (and age) of the trees involved
- surges of naturally produced acids
- acid flushes (e.g. during spring snow melt or after prolonged drought).

There is growing interest in the possibility that ozone might be a more likely cause of tree damage than acid deposition, through increasing the vulnerability of individual trees to poisoning and loss of nutrients (Skeffington and Roberts 1985; *The Times*, 8 May 1986). Moreover, some research suggests that increased ozone makes trees more vulnerable to acid-induced damage. Ozone might play a significant role, especially at high altitudes where sunshine (required for the photo-chemical processes that produce the ozone) is more intense. Many of the damaged trees in West Germany and elsewhere are found above the 10,000-foot line, in remote areas of forest some way away from obvious centres of pollution (Ashmore, Bell, and Rutter 1985; Tifft 1985: 47), suggesting that ozone may be not only involved but perhaps of some significance.

Puzzles and problems

Many key problems and areas of uncertainty remain as yet only partially resolved. Part of the variability of response arises because much damage may remain undetected or invisible until it reaches a critical, perhaps irreversible stage. Different species have different dose-response relationships, so they respond to changing levels of acid input in different ways.

Further variability arises because damage to vegetation tends to be species specific, so that different plant species react to a given level of acidification in different ways. Furthermore, not all individuals of a given species are equally sensitive to pollution stress. Genetic variability gives rise to variability in response to SO_2 and ozone in populations of Scots pine, Norway spruce, and other coniferous (and some species of deciduous) trees, for example (Lesinski 1983).

Moreover, much damage to vegetation is location specific (Buckley-Golder 1985). Thus one forest might show signs of damage from acid deposition whilst a neighbouring one (perhaps underlain by different soils or bedrock, with different buffer capacities) might not.

To complicate matters even further, synergistic (compound) effects (see chapter 3) are often apparent. Mixtures of SO_2, NO_x and O_3 occurring together appear to have a far greater effect than each gas acting on its own (Buckley-Golder 1985).

One particular puzzle is why lichens, which are very intolerant of SO_2 pollution, still exist in some woodlands where tree damage and death have been attributed to acid deposition. The reverse side of the same coin is the puzzling lack of reports of tree damage and death in heavily polluted city centres from which lichens have long since disappeared (Benarie 1985). This again suggests that ozone may play a more critical role than some scientists have hitherto thought likely.

'Acid rain' is a strange and variable brew of pollutants (see chapter 2), and the variable composition/character of the mixture through time and from place to place adds a further veneer of complexity to an already highly variable family of responses by vegetation. The net effect of all these problems is extreme difficulty in disentangling possible causal or contributory factors. The air of mystery persists.

CROP DAMAGE

We have concentrated in this chapter on effects of acidification on trees and forests. This focus is appropriate, given the evidence of serious damage in many areas and the correspondingly high interest of conservationists in the problem. However, we should not overlook the possible effects of acidification on other forms of vegetation, most especially on agricultural crops.

Sulphur dioxide and wet deposition of acids affect all higher plants, either in reduced growth or in physical damage (both result in reduced crop yields). It has been estimated (Elsworth 1984a) that acid rain has destroyed up to £25 million worth of crops in parts of Cumbria and Scotland – the areas of Britain that now receive the highest atmospheric inputs of acids (see chapter 1). Fears have also been expressed that acid rain might be cutting crop yields in Britain by as much as 10 per cent, costing the nation's farmers in the region of £200 million each year (M. Unsworth, quoted in *Acid News*, 1984, 2). Farmers meet part of the cost of acid rain via increased need to add lime to acid soils, as well as through reduced crop and livestock yields (Figure 5.6).

As yet, however, there is no *conclusive* evidence of actual crop damage

Figure 5.6   Effects of acid rain on farming

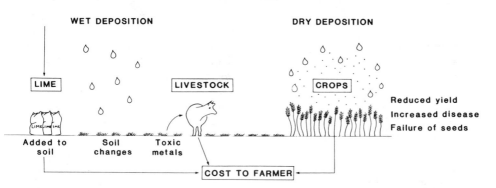

caused by acid rain, and laboratory tests have yielded mixed results: some have shown reductions in crop yield, some have shown increases, and some have shown no net change (La Bastille 1981: 672–3; Roberts *et al.* 1983; Roberts 1984). Part of the variability might stem from the fact that sulphur and nitrogen can act as plant nutrients, rather than damaging pollutants. Below critical thresholds and in the short term, acid deposition can be an asset in providing limiting nutrients (especially nitrogen) and hence promoting plant growth (increasing forest productivity). Laboratory studies have shown that, at low concentrations, SO_2 pollution (or fumigation) stimulates the assimilation of CO_2 by plants, and thus it assists plant growth (Lesinski 1983).

Most damage appears to stem from quite subtle indirect effects of wet and dry deposition. The two primary causes of indirect effects are the changing amounts of available plant nutrients in the plant and soil system, and the mobilization of compounds (e.g. heavy metals) that are toxic to plants. Both involve interactions between plant and soil, so that acidification of soils is of critical importance.

TREES AS POLLUTION FILTERS

The association between plants and air pollution is symbiotic (two-way). Air pollution affects plants, directly and indirectly, generally for the worse. But plants also affect air pollution, generally for the better. Trees, in particular, operate as natural air filters that remove (scavenge) pollutants from the atmosphere and thus improve air quality. Forests act as 'pollution sinks' in two main ways – as air filters (grime that collects on the leaves of trees is removed from circulation in the atmosphere) and as air ventilators (trees cause air currents and eddies that help to ventilate an area that might otherwise have very still air) (Mabey 1974: 99).

Coniferous trees are particularly effective in filtering out gaseous pollutants from overying air (Harriman and Morrison 1981; Last and

Nicholson 1982), although this is a rather mixed blessing. The good news is that the trees act as filters, scavenging noxious particles from the overlying air. The bad news is that the trees and underlying soils absorb the toxic materials that are scavenged. This means that the trees themselves can increase the acidity of rainfall (i.e. enrich it) within the forest. After heavy rain the sulphur particles that have collected (dry) on conifer needles are washed off on to underlying soils and into adjacent streams and lakes. Coniferous trees can also intercept acid mists (occult precipitation – see chapter 2) that might otherwise have passed over an area without allowing deposition of acids, a process that can further enhance the acidity of throughfall beneath the forest canopy (Buckley-Golder 1985).

Because of this symbiotic link between trees and pollution, there are quite compelling environmental grounds for concluding that a policy of planting conifers in Britain's marginal uplands (see chapter 3) must be seen very much as a mixed blessing.

CONCLUSIONS

As with surface waters (chapter 4), we emerge with the conclusion that, whilst it is difficult to find unequivocal evidence for or against the proposition that acid deposition causes unacceptable damage to soils, trees, and crops, the circumstantial evidence is strong and convincing.

The possible impacts are both biological and economic. Biological effects are reflected in tree dieback, reduced productivity, falling crop yields, contamination of crops with heavy metals, and so on. But the economic dimension is also very real. The decline in tree productivity is directly translated into a reduced supply of timber (hence income from forestry), and the decline in crop yields is a serious commercial blow for the farmer. Both also produce market shortages that encourage price rises, so that consumers also pay the cost of acidification.

Many uncertainties surround any evaluation of the effects of acid rain on forests and soils, not least of which is the fact that the biological effects of acidification change as acid concentration increases. At low concentrations, dry and/or wet fallout can provide valuable free nutrients to plants, especially in nutrient-deficient soils. But at higher concentrations (and the critical threshold varies from species to species, and from area to area depending on buffer capacities) direct damage to plants might be expected. This can take various forms, from reduced rates of growth to severe physical distortion and disfigurement (especially in trees). Added complexity comes from the fact that reported damage, to trees in particular, is widely scattered – even within one forest. Damage is *not* ubiquitous. This might imply that different explanations are required for different areas. Indeed, some scientists stress that it is as difficult to explain why acid deposition has *not* affected some areas, as it is to explain why it *has* affected others.

Many scientists have warned that unless drastic action is taken without delay, large tracts of Europe's forests may be reduced to barren wastes by the turn of the century (Halford 1986). Wholesale loss of forests would threaten the forestry and timber industries in Europe, which provide much-needed raw materials and many jobs. Tourism could also be affected, with implications for employment and local economies.

The sort of action that is being widely called for centres on reducing emissions (mainly of SO_2, but also NO_x) from power stations, factories, houses, and vehicles in order to reduce the potential for producing acid deposition. As we shall see in the next chapter, similar calls are also being made by those who are concerned about the possible damage caused by acid rain to buildings and to humans.

6 EFFECTS ON BUILDINGS AND HUMANS

Palaces and people . . .

It is a reverend thing to see an ancient castle or building not in decay.
(Francis Bacon, *The Elements of Common Law*,
Volume 14, *Of Nobility*)

Most of the immediate controversy surrounding the impacts of acid rain
has centred on lakes and rivers, and on forests and crops. This focus of
concern is easy to understand, given the mounting evidence of biological
damage in Scandinavia, West Germany, Britain, Canada, and the United
States. Whilst the natural history and conservation interest has over-
shadowed scientific concern over other unwanted impacts, at least in terms
of media coverage and general public awareness, the latter are proving to
be both real and significant. They are also proving to be costly. A recent
OECD estimate put the cost of damage to buildings in Western Europe
from industrial air pollution at around $3.5 billion per year, and the cost to
human health at a further $8 billion per year (*The Times*, 19 November
1985, p. 16).

Impacts on buildings and humans have *not* (or perhaps *not yet*) figured
more prominently in the 'acid rain debate' largely because they affect
smaller areas (large cities and industrial areas) and the damage appears to
be related more to dry deposition (oxides, gases, and particles) than to wet
deposition. Most concern in the debate has focused on the long-range
(especially the trans-national) effects of wet deposition – acid rain in the
true sense.

Forests and lakes are being affected by acidification in the countryside;
buildings and structures are being affected in the cities. We shall examine
effects on materials and buildings and effects on human health in this
chapter.

EFFECTS ON MATERIALS AND BUILDINGS

Air pollution attacks stone, metals, and fabrics, and there is no shortage of
evidence of the damaging effects of many forms of air pollution on
buildings and structures. Acid deposition (wet and dry) is widely
implicated as the most serious cause of the damage.

The problem is vividly illustrated in Cracow, Poland's major city, regarded as having one of the world's worst pollution problems, which receives acid deposition from upwind Katowice (a heavy industrial area); 'as a result Cracow is crumbling. The golden roof of the church had to be removed because it was dissolving. The faces of the statues are melting. Steeples are falling down, balconies disintegrating' (May 1982: 9).

Perspective on the problem

Urban air pollution is *not* a new phenomenon. In 1661, for example, the noted philanthropist John Evelyn submitted his text 'Fumifugium, or the Smoake of London Dissipated' to King Charles II. He wrote of early seventeenth-century London,

> whilst these [chimneys] are belching forth their sooty jaws, the city of London resembles the face rather of Mount Etna . . . . or the suburbs of hell, than an assembly of rational creatures, and the Imperial Seat of our Incomparable Monarch . . . . The weary traveller, at many miles distance, sooner smells than sees the city to which he repairs. This is that pernicious smoke which foils all her Glory, superinducing a sooty crust or fur upon all that it lights.   (quoted in Thompson 1978: 6)

Neither is the problem of building damage from air pollution new, or newly discovered. As far back as 1852, Robert Smith (the grandfather of acid rain studies, mentioned in chapter 1) reported to the Chemical Society:

> it has often been observed that the stones and bricks of buildings, especially under projecting parts, crumble more readily in large towns, where much coal is burnt, then elsewhere. Although this is not sufficient to prove an evil of the highest magnitude, it is still worthy of observation, first as a fact, and next as affecting the value of property. I was led to attribute this effect to the slow but constant action of the acid rain. (Smith 1852)

What *is* new is the growing seriousness of the problem. As SO_2 concentrations in the atmosphere continue to rise, and acid deposition becomes a greater and more widespread problem, damage to buildings and materials gets worse and more widespread.

Historic materials offer a valuable basis for charting the progress of damage through time. For example the Elgin Marbles, which were transported from the Acropolis in Rome to the British Museum in London in the late eighteenth century, are well preserved in the clean museum air, whereas sculptures of a similar age that remained *in situ* in Rome have since deteriorated badly in the polluted urban air (McCormick 1985). It is generally believed that air pollution has caused more deterioration of ancient statues in Rome since 1950 than had occurred in the previous 2,000 years (Gates 1972).

Corrosion of materials

Air pollution speeds the natural chemical weathering and corrosion of exposed materials in a variety of ways. For example, ferrous metals are attacked by sulphur oxides, iron rusts more quickly, and zinc products are more badly corroded in urban areas. Sulphur oxides etch the surface of the metals, and when iron rusts the surface becomes flaky and the flakes fall off to expose more metal to the etching process. Consequently steel buildings, railway rails and other structures built of iron can be very seriously affected by air pollution, with extensive economic losses (Lynn 1976; Anon 1984/5d). Non-ferrous metals corrode much less, because the oxides and sulphates remain on the surface and offer protection to the metal below.

Experimental tests show that many materials corrode faster in polluted urban and industrial air (particularly in the presence of high concentrations of sulphur dioxide) than they do in the countryside: corrosion rates are commonly between two and ten times as fast (Tolba 1983: 118). Corrosion rates are often speeded up even more in the presence of moisture and carbon soot and with high temperatures; hence corrosion tests in Sweden have shown that carbon steel is corroded fifteen times faster in Stockholm's polluted urban air than in a rural sub-arctic area (with low temperatures and low SO_2 and sulphate levels) (Perkins 1974). Corrosion tests tend to show only marginal effects of air pollution on materials like aluminium and stainless steel.

Thus air pollution speeds rates of deterioration and accelerates the spread of damage in materials like carbon steel (coated and uncoated), painted steel, galvanized steel, copper, nickel, and nickel-plated steel. Consequently items made of these materials are at risk from acid deposition, particularly in the dry form (gases, oxides, and particles of sulphur). Thus, for example, cars and other road vehicles are believed to corrode faster in major cities than in the countryside (Greater London Council 1985: 17), with corresponding economic implications for their owners! Similarly, electrical devices containing copper, silver, and gold contacts are more at risk from corrosion in heavily polluted areas; silver used in electrical contacts in telephone exchanges often has to be protected from tarnishing (Greater London Council 1985: 17).

Sulphur oxides are only *one* component of air pollution in many cities; other components can also produce damaging effects. For instance, ozone has long been known to cause damage to rubber, mainly by increasing cracking and thus speeding deterioration. There is a belief that the high levels of oxidants and ozone in the air of Los Angeles (caused by exhaust emissions) seriously reduces the life of rubber tyres (Strauss and Mainwaring 1984). The vehicle owners also pay for air pollution in increased running costs! It has also been found that, when rubber tyres were stacked in warehouses in Los Angeles, the tyres on the bottom often cracked (by ground-level ozone concentrations) during storage (Carr 1965).

Air pollution can also damage paint coatings. Sunlight normally causes most direct damage to paintwork, but rates of deterioration speed up considerably in the presence of sulphur dioxide, ozone, and particulates (Greater London Council 1985). This not only spoils visible appearances through discolouration, it fails to provide proper weather proofing so that underlying materials are then directly exposed to the ravages of weather and pollution. There is some evidence to suggest that intervals between repainting tend to be shorter in urban areas than in the countryside (Greater London Council 1985), but this might not just reflect observed differences in air quality.

If surface and ground waters are affected by acidification, submerged structures may be corroded. This might affect a much wider area than the immediate vicinity of urban and industrial centres. Thus bridges, dams, industrial equipment, water supply networks, underground storage tanks, and hydro-electric turbines over a wide area can also be seriously affected by acid deposition (Tolba 1983).

Damage to buildings

The greatest concern over effects of acid rain on materials has centred on damage to buildings, especially those built of sandstone, limestone, and marble. Rates of decay and damage to stone in urban areas are often two to three times as high as in rural areas (Winkler 1970) because of accelerated weathering from air pollution. Both the stone and the mortar of buildings can be affected, and dry deposition appears to be the principal damaging agent. Pollutants are normally of *local* origin.

Limestone buildings and structures are normally worst affected. Dry deposited sulphur reacts with the calcium carbonate in the presence of moisture to form calcium sulphate. This is soluble, and the acids so formed are washed off the surface of the stone when it next rains (Bottam 1966; Greater London Council 1985). This cycle of dry deposition–reaction–washing is repeated many times, and greatly speeds the natural weathering and erosion of the building fabric. But there is also another process at work (Fassina 1978). The soluble sulphates, nitrates, chloride and other salts that are washed across the surface of the stonework will crystallize within the stone when the water evaporates off. As they turn into crystals, the salts expand and this force enlarges the cracks; this leads to a crumbling (exfoliation) of the stone surface, which in turn exposes unweathered stone beneath to the cycle of dry deposition–reaction–washing (Dunmore 1986). In winter, moisture can also penetrate the cracks, and then freeze, expand, and cause further flaking. Chemical corrosion causes all types of limestone to powder, crumble, flake, crack, or chip. The direct roles in such erosion played by wet acid deposition and by dry deposition of nitrogen oxides are as yet *not* fully understood or quantified.

Different building materials are affected to different extents. Worst affected are limestones and dolomite. But since (as we shall see in chapter 7) one of the proposed clean-up techniques for SO_2 in power station emissions is to react it with lime, it should come as *no* surprise to find that the same chemical reaction takes place in nature (Perkins 1974)! Similarly cement (which has a high lime content) can be seriously weakened by SO_2. In contrast, sandstone, which is composed mainly of silica, is relatively unaffected by SO_2; the main visible effect is to produce a hard, black surface coating on exposed sandstones. Granite also blackens rather than crumbles.

One common effect of acid deposition is to discolour building materials and paintwork. This staining by sulphur pollutants is especially visible in industrial cities where initially light-coloured stones and bricks soon become dark and blackened. Such 'damage' has a clear economic implication via the need for regular and costly cleaning, repair, and renovation works.

Brick-built structures are less vulnerable to air pollution damage than stone-built structures. As a result, relatively few recent (brick-built) buildings face serious erosion problems, but a great many (stone-built) historic monuments and buildings do. Some of Europe's finest Gothic (twelfth to sixteenth century) cathedrals are showing such serious signs of damage and deterioration that some experts fear that – without marked reductions in local urban air pollution – they might not survive for more than about two more centuries (Strauss and Mainwaring 1984). There is little doubt that most of the damage is recent: 'comparisons with old records still in existence reveal that the 150 years since the beginning of the Industrial Revolution have caused more damage than occurred in all the previous centuries' (Lynn 1976: 145).

In some countries (especially Britain and the United States) it is believed that rates of erosion of building materials may have slowed down in recent years, reflecting reductions in SO_2 levels because of Clean Air Acts. But there *are* signs that there may be delays between emission and damage, thus delays might be expected between reduced emissions and reduced erosion. It has been suggested that most stone deterioration now evident is mainly the result of dry deposition of SO_2 from the years before the Clean Air Acts were passed (the early 1960s), rather than from current industrial emissions (Ridley 1985). This suggests that we should perhaps be thinking of an 'acid time bomb' – the effects of which might lie dormant for some time, accumulating to a serious level before visible evidence starts to appear.

Erosion of the cultural heritage

There are growing concerns about the extent of visible damage now being caused by air pollution to the world's cultural heritage – ancient monuments, historic buildings, sculptures, ornaments, and other important cultural objects. The catalogue of significant casualties grows longer each year. The following discussion is *not* intended to be exhaustive; it offers a selection of the best (or, to be more correct, worst) examples.

Many of Europe's most cherished cultural treasures are housed in Athens in Greece. Yet all is not well there. Acid precipitation (wet and dry) has increased dramatically over the present century with the increased emission of oxides from industry, vehicles, and domestic heating systems. Many of the city's most famous artefacts are paying the toll. The stonework of the Acropolis is believed to have crumbled more during the past forty years than it had done over the previous 2,500 years (Strauss and Mainwaring 1984). The Parthenon has suffered so much damage from air pollution that some official reports predict it might fall down completely without extensive (and costly) repair work (Thomson 1969).

Italy vies with Greece as a cultural centre; it is also faced with similar problems from air pollution. In Rome, Trajan's Column and Michelangelo's Statue of Emperor Marcus Aurelius have been badly damaged (the latter had to be removed because of metal corrosion) (May 1982: 9). Many of the finest palaces and monuments in Venice are also being attacked by dry deposition of pollutants, particularly 'oil-fired [acid] carbonaceous particles [which] are among the main agents responsible for stone weathering' (Camuffo, Moule, and Ongaro 1984: 135). In many of Italy's major cities, sulphur dioxide has damaged frescos made of pigmented lime plaster; the plaster is changed to gypsum, and in the process it shrinks in volume and loses its coherence (Thomson 1969). Crumbling fragments are all that remain of some of the country's finest artistic and cultural heritage; *no* amount of money can replace them.

Damage is not restricted to major cultural or tourist centres. In the Netherlands, for example, air pollution has caused serious damage to statues and other decorations outside the 457-year-old's Hertogenbosch Saint John's Cathedral. Amsterdam's Royal Palace on Dam Square and the nearby Rijksmuseum are also suffering from black encrustations ('*stone cancer*') caused by SO_2 in the air (Hoyle 1982). Sweden, already incensed at the loss of fish stocks in many rivers from acidification arguably imported from abroad (see chapter 4), is also faced with the ignominy of producing its own urban pollution that affects buildings and treasures. Stockholm's thirteenth-century Riddarholm Church had a cast-iron spire that was so badly corroded by acid rain that it had to be replaced in the 1960s; the church's sandstone Royal Burial Chapel is now being seriously damaged by air pollution (Hoyle 1982).

Neither is damage confined to Europe. In India, the Taj Mahal has been a 'jewel in the crown' of grand architecture and a valuable tourist attraction. But sulphur pollution from power stations at Agra and a large oil refinery at Malthura has brought many problems to the building, including discolouration, flaking of sandstone, and blistering of many of the precious stones set in to the walls (May 1982: 9). Delhi's Red Fort and Jama Masjid are also showing signs of damage from sulphur pollution, in a city that has experienced a 75 per cent increase in air pollution since the mid-1970s (Dutt 1984).

The United States, a country with a brief if rather eventful 'history', none the less has its fair share of damage to buildings and structures. Cleopatra's Needle, a grand stone obelisk transported from Egypt to New York in the 1860s, provides a measure of how fast pollution-accelerated damage can be. The column was originally erected by the Nile opposite Cairo, around 1500 BC; all of its inscriptions were in good condition when it left Egypt. Today, legible hieroglyphic inscriptions survive on the east side of the column, but the west side (facing prevailing winds) has been eroded almost flat (May 1982: 9). This simple visible evidence shows that ninety years in New York have done more damage to the monument than 3,500 years in Egypt had done. New York's famed Statue of Liberty – admittedly metal rather than stone – celebrated its hundredth birthday in July 1986 (Stengel 1986) only after a massive (and costly) repair and renovation job.

London, England, has long been one of the world's cultural centres, and – as noted in chapter 1 – it has long produced and suffered from air pollution. It has also provided some monumental examples of building damage. The Houses of Parliament need regular repairs to the stonework, partly because of damage caused by air pollution (Ormerod 1983); external repairs carried out in 1983/4 alone cost £624,000. Also in London, one of four bronze statues had to be removed from the Royal Artillery Memorial (completed in 1925) at Hyde Park Corner late in 1984 for cleaning, at a cost of around £30,000, because acid rain had corroded its iron and steel fixings and was destroying the supportive relief carvings in soft Portland stone (*The Times*, 10 October 1984).

Christopher Wren started to build St Paul's Cathedral in 1675. But in 1717 a decision was taken to add a balustrade, and this was done the next year. When the balustrade was built, the joints between and lifting holes within the Portland limestone blocks were filled (flush with the stone surface) with lead. The stonework has been eroded by air pollution, the lead has *not*, so the difference in height between stone and lead gives a measure of rates of stone erosion. Between 1718 and 1980 the stone was being worn down at a rate of around 8 mm per 100 years (around 20 mm over the 250 years). Detailed measurements of present-day rates are in the range 5.8–31.9 mm per 100 years (Sharp *et al.* 1982), and over two-thirds of the estimated total stone loss on exposed parts (such as the South Portico) is believed to have occurred within the last forty-five years. Stone weathering and discolouration are costly to repair: cleaning and renovation of the Cathedral's North-West Tower during the early 1980s cost in the region of £300,000 (Friends of the Earth 1985).

Elsewhere in London, Westminster Abbey has also suffered extensive damage from air pollution. Restoration of external stonework began in 1973, and the plan is to reface almost all of the Abbey surface with Portland stone, at a cost in excess of £10 million. Completion is not expected before the early 1990s (*The Times*, 20 September 1985).

St Paul's and Westminster are not the only British cathedrals facing serious problems from air pollution. The UK House of Commons Select Committee (see chapter 10) noted that many of Britain's finest grand buildings are being affected (House of Commons Environment Committee 1984) – including Lincoln Cathedral, York Minster, Beverley Minster, Brompton Oratory, Liverpool Cathedral, and the Palace of Westminster. Friends of the Earth (1985) point out that sulphur from central heating boilers may be important in city centres, but power station emissions are most likely sources at rural sites (e.g. Lincoln Cathedral). Salisbury Cathedral, painted by Constable and visited by millions, is also on the casualty list. Its famous spire – which has stood tall for six and a half centuries – is now endangered by crumbling stonework and corrosion damage and in 1985 an appeal was launched to raise the estimated £6.5 million cost of restoring the spire and its supporting tower (Spence 1985).

Cultural treasures

Stained glass windows are also badly affected by air pollution. A world-famous specialist in the conservation and restoration of glass concluded in late 1985 that 'if stained glass windows are kept in situ in their present state of preservation, their total ruin can be predicted within our generation' (Frenzel 1985: 100) simply because of the corrosive effects of air pollution. He gives the example of West Germany's massive, twin-spired, high Gothic Cologne Cathedral (built around AD 1200), which was heavily adorned with grand stained glass windows. Continuous etching by air pollution has corroded the exterior surface of the glass, reducing its thickness year by year, and giving the decomposed surface a 'weathering crust'. Rain washes over the crust and begins the cycle of destruction again. Meanwhile the coloured glass itself breaks into tiny particles, which fall out of the panel. This relentless attack by air pollution is not confined to Cologne – many of Europe's finest cathedrals (including Canterbury in England and Chartres in France) are suffering in a similar way. Over 100,000 stained glass objects in Europe are believed to be threatened, with glass from between the eighth and seventeenth centuries at greatest risk (*The Times*, 29 September 1984).

Damage to the cultural heritage from air pollution is *not* confined to major structures. Neither is it confined to objects exposed out of doors. The air inside art galleries, museums, libraries, churches, and other buildings also contains air pollutants (derived from outside), and the treasures housed within them – paintings, drawings, fabrics, antique costumes, and so on – have suffered accordingly. Whilst many of the world's most important repositories are now air-conditioned to aid preservation, it is known that dry particles of sulphur dioxide can infiltrate simple ventilation systems and thus threaten contents (Hoyle 1982).

Fabrics soil, fade, and lose their tensile strength (i.e. rip more easily) in the presence of air pollution; regular cleaning and renovation are required (at a cost). Valuable oil paintings in non-air-conditioned museums have been found to be damaged by air pollutants, requiring careful (and costly) restoration and conservation work (Stern *et al.* 1973). Leather becomes brittle when exposed to sulphur products, and as far back as 1843 Faraday concluded that the rotting of leather upholstery on chairs in a London gentleman's club was being caused by sulphur compounds in the air (Stern *et al.* 1973). Leather bindings on old books are also affected, and repeated opening of the embrittled spines can cause considerable damage.

Old papers and books are also threatened because paper absorbs SO_2 and converts it to sulphuric acid (H_2SO_4), which is damaging. Damage is confined mainly to paper made after 1750, and particularly to modern paper made by chemical processes (metals in the processed paper speed up the conversion of SO_2 to sulphuric acid; older paper was made from natural materials and lacks chemical impurities) – the paper becomes very brittle and loses its fold resistance (Carr 1965). Thus old books that are not stored in sealed cases undergo gradual deterioration.

Cost of damage

This review has had to be selective, but it does illustrate the many ways in which acid deposition (normally in the dry form) affects materials. As noted earlier, this dimension of the acid rain debate has attracted less attention than possible effects on forests and lakes, but it is serious and it does cost money. The former Greater London Council concluded in 1985 that 'deterioration of materials and buildings is probably the most costly (in economic terms) aspect of air pollution damage' (p. 17). It is shared widely and difficult to distinguish from normal routine building repair and maintenance costs.

Damage to materials is amongst the most easily quantifiable effects of acid deposition – decreased lifetime, replacement value, or restoration costs can often be calculated or estimated with reasonable certainty (Whelpdale 1983). Various estimates have been offered of the cost of damage to buildings and materials from air pollution (Stern *et al.* 1973) – e.g. total costs in the USA (late 1960s) $104 billion per annum; total costs in Great Britain (early 1950s) $1.4 billion per annum; total costs in France (early 1960s) $0.6 billion per annum.

Much of the cost of damage to materials arises from corrosion and erosion. A 1981 Economic Commission for Europe study (Tolba 1983: 118) found that corrosion of painted and galvanized steel structures costs $2–10 per year for each person living in Europe. These figures are of a comparable order of magnitude to earlier estimates – e.g. of material corrosion in the United States (1970) at $7.10 per person per year, in Sweden (1970) at $4.30 per person per year (Kucera 1976).

Some pioneer attempts to estimate costs were made in the Beaver Committee Report in 1954 (see chapter 1). It was estimated that total economic loss from air pollution in England was around £250 million per year (at mid-1950s' prices), comprising £150 million direct costs and a further £100 million indirect costs (related to loss of efficiency – e.g. in operation of transportation facilities). The £150 million direct costs break down, roughly, as follows: laundry (£25m.), painting and decorating (£30m.), cleaning and depreciation of buildings other than houses (£20m.), corrosion of metals (£25m.), damage to textiles and other goods (c. £50m.).

It has recently been estimated (Greater London Council 1985) that air pollution produces damage to buildings, materials, and structures across the United Kingdom as a whole in the range £16–£770 million per year (£0.28–13.75 per person per year across the UK); and damage to buildings, etc. in London of £2–100 million per year (i.e. £0.30–14.90 per person per year in London).

It is difficult to give precise estimates of the true cost of building damage, especially given the problems of defining the value of historic buildings and treasures. It is clear that 'no amount of research can produce a costing for heritage loss caused by the damage to historic monuments and statuary . . . when such national and international treasures are damaged beyond redemption, the world loses a priceless asset, for all time' (Dunmore 1986: 116). Accepting this difficulty, most estimates are based simply on repair costs. Thus, for example, acid rain is believed to cause around $10 million damage each year to historic monuments in the Netherlands; damage to Cologne Cathedral in West Germany costs $2 million a year simply to restore (Hoyle 1982).

It would be unwise to place too much emphasis on estimates of economic costs, because of the difficulties of arriving at realistic and meaningful estimates. But however the figures are arrived at, we are talking about vast sums of money! None the less, this is only *one* component of the general problem, and we must turn now to possible effects of acid deposition on human health.

EFFECTS ON HUMAN HEALTH

An evaluation of the problems caused by acid deposition would be incomplete without giving some consideration to possible effects on human health. This theme has not attracted anything like such wide public interest as damage to forests and lakes, but this is understandable given that (unlike many forms of pollution) acid rain cannot be felt, smelt, tasted, or seen. Acid rain itself is a truly invisible form of pollution. Moreover, acid rain is not known to be mutagenic (i.e. gene changing) or carcinogenic (i.e. cancer forming) in humans. No one is known to have died directly from acid rain, and there are – as yet – few clear direct links with human diseases and disorders.

Despite this lack of strong public interest, the issue is not without significance. There is a spectrum of effects of acid deposition on humans, ranging from mere nuisance (such as bad smells, reduced visibility, irritation of the eyes, nose, and throat) through to increased mortality (Crowe 1968). There is now firm evidence that air pollution and human health are closely associated in space and time. For example, the region of Katowice in southern Poland (source area of the pollution that falls on Cracow and damages buildings there) suffers 47 per cent more respiratory disease, 30 per cent more tumours, and 15 per cent more circulatory disease than the rest of Poland (May 1982: 10). Furthermore, 'although there are no official figures collated as yet [1984], there are signs that babies have actually died in certain areas of West Germany from a throat infection, and the incidence of death is highest where there is a greater direct pollution problem' (*The Times*, 25 September 1984, p. 4). Death rates from bronchial problems in London have been much higher in the past during short episodes of severe air pollution (see pages 126–7).

It is perhaps convenient to distinguish two ways in which air pollution might affect human health – directly and indirectly. Direct effects involve damage to the human body from exposure to pollutants, and in this sense most damage is caused by sulphur dioxide in the dry gaseous and aerosol form. Acid rain *per se* (i.e. wet deposition) has no *known* direct effects on human health. Indirect effects involve damage to humans by contact with materials that have themselves been affected by acidification (by wet or dry deposition) – obvious examples are food and water supplies.

Direct effects

As noted above, direct effects arise largely from the fact that SO_2 is a highly poisonous gas. It is described as a 'non-specific irritant', which remains outside the body's biochemistry; the body reacts to its presence (as opposed to 'systemic poisons' like lead, which affect specific body processes) (Lynn 1976: 83). In all higher animals (including humans), SO_2 is taken in by inhalation and attacks the respiratory system (mainly the throat and lungs). In humans it is associated with chronic bronchitis, pulmonary emphysema (constriction of the lungs, which makes breathing difficult), and – it is believed – cancer of the lungs.

Dose–response relationships

The greater the concentration of SO_2 in the air breathed, the greater the effects on human health. But the toxicity of SO_2 varies with concentration, and it is difficult to establish a 'safe level'. Laboratory studies have established that the dose–response relationships for SO_2, NO_x, and sulphate particles are *not* linear (i.e. a doubling of concentration produces more than a doubling of the damage), and that there are threshold values below which effects appear to be minimal. The effects of low concentrations (i.e. low doses) are generally small and always difficult to detect.

Most damage to human health is caused by SO_2 in gas and aerosol forms. Concentrations below about 0.6 ppm (parts of SO_2 per million parts of air) normally produce no detectable responses in healthy human beings (Air Conservation Commission 1965). It is known that a concentration of 1.6

ppm will cause a temporary (and reversible) tightening of the bronchi (Duffus 1980) – the tiny passages in the lung via which oxygen (and contaminants like SO_2) are absorbed into the bloodstream. Above 1.6 ppm, breathing becomes detectably more difficult and eye irritation increases (because the SO_2 readily dissolves in the fluids that coat the human eye). Background levels commonly found in the air of industrial areas and major cities are in the order of 1 ppm (but levels can be much higher for short periods). But those who work in the chemical industry often face occupational exposure to SO_2 concentrations in the range 1–5 ppm (Air Conservation Commission 1965). Most people can detect 5 ppm (exposure for one hour causes choking), and they find 10 ppm unpleasant (exposure for one hour causes severe stress from breathlessness and tightness of the chest). Sulphur dioxide increases the risk of respiratory infection amongst those who already suffer from respiratory deficiencies (e.g. asthma sufferers). Prolonged exposure to high levels of SO_2 *can* cause death in humans, from respiratory problems or heart failure (when breathing and any form of physical activity become so difficult that they put excess strain on the heart).

Nitrogen oxides are much less active biologically, so they cause fewer problems for human health and much higher concentrations can normally be tolerated. Studies of occupational exposure show that most people exhibit only weak signs of lung disorders (pulmonary fibrosis and emphysema) at concentrations in the range 10–40 ppm (Lynn 1976: 83). High concentrations of nitrogen dioxide (which interacts with some hydrocarbons and sunlight to produce complex chemical irritants) are also known to cause eye irritation. Normal atmospheric levels in cities and industrial areas rarely exceed 0.5 ppm, so health is rarely affected.

It is difficult to draw generalizations about how people respond to a given level of SO_2 or NO_x because some individuals can tolerate much higher levels than others. Susceptibility varies considerably between individuals; those most at risk are the aged, those who suffer from heart diseases, those with impaired lung capacity, and those who are hyper-sensitive (perhaps because of earlier disease).

Generalizations are also difficult to arrive at because health is most badly affected by synergistic reactions (i.e. complex chemical interactions that bring worse problems than the individual chemicals bring on their own). SO_2 is much more toxic and damaging when combined with aerosols, acid mists, and suspended smoke (Lynn 1976: 83), because these chemical cocktails form finer suspensions that penetrate the lungs further than the gas alone. Moreover, in smoky conditions we tend to breathe more heavily, and so – paradoxically – ensure greater penetration of the lungs by the pollutants.

Most of the non-terminal effects cause short-lived suffering to the people who are exposed to the temporarily high atmospheric levels of SO_2 (e.g. associated with a bad episode of pollution). But because the problems are rarely so acute that hospitalization or visits to doctors are required, it is difficult to establish just how widespread or serious they are. One indicator is the number of complaints lodged with the authorities; for example, over 32,000 complaints were received about high sulphur levels around Tokyo in July 1974, and on numerous occasions since then people cycling or working out of doors in Tokyo have complained of suffering from eye and skin irritations believed to be caused by polluted drizzle droplets (Okita 1983: 101).

Disastrous pollution episodes

It is perhaps understandable that the best evidence is available for major disasters involving large-scale death, where the association between air pollution and death is stronger than circumstantial. Three such events involving smogs are well documented (Meetham 1964; Perkins 1974).

The first began on 1 December 1930 in the Meuse Valley near Liège in Belgium (Firket 1936). This is a heavy industrial area, with many industrial point sources of pollution (especially SO_2), including iron and steel works, zinc works, glass works, potteries, lime kilns, electricity-generating stations, and sulphuric acid plants. A thermal inversion (cold air overlying warm) trapped pollutants in the low-lying valley, and the lack of natural ventilation in the stagnant air meant that pollutant concentrations rose over the five days of thick fog. The factories produced a cocktail of pollutants (including SO_2, sulphuric acid, and zinc aluminium sulphate), which caused severe respiratory problems for several hundred people in the area. But the cocktail was a deadly one: sixty-three people died, mainly on 4 and 5 December and mainly by heart failure after prolonged periods of acute irritation of the respiratory system and vomiting. SO_2 concentrations were later estimated to be in the range 9–38 ppm, which – if true – is exceedingly high.

Similar weather conditions gave rise to the disaster that befell the town of Donora, south of Pittsburgh, USA, in late October 1948 (Schrenk *et al.* 1949). A fog closed over the town on the 26th, with air trapped close to the ground by thermal inversion. When it lifted five days later, it left a death toll of twenty (all were over 50 years old, most died on the third day). Nearly half of the town's population (of 14,100) had been made ill, and a tenth had been severely affected; symptoms included irritation of the eyes, nose, and throat, coughing, respiratory irritation, headache, and vomiting. The cause of the disaster was sulphur pollution from the town's steel mill, zinc production plant, and sulphuric acid plant. Again, no measurements of SO_2 levels were made at the time of the disaster, but later estimates put the concentration at 0.5–2 ppm.

These two events were insignificant in comparison with the 'killer smog' that hung over London, England, between 5 and 9 December 1952 (Wilkins 1954; Wise 1971). It started, again by inversion, on the Thursday. A white fog formed, which was soon to become black, mainly from smoke (coal was used heavily in heating houses and in local power stations). This was yet another 'pea-souper' for which London had become notorious (see chapter 1). The fog became thicker and pollutants built up in the stagnant air. Visibility soon fell to almost zero. One Londoner gave a vivid description: 'you couldn't see one's hand in front of one's face – a white shirt collar became almost black within 20 minutes – smog was intensely irritating to eyes, throat and bronchi. A cough soon developed. The cough was to remain with me for months' (Perkins 1974: 325). He was lucky . . . the smog took an estimated 4,000 lives, mainly from breathing problems. The deaths were spread over a period of weeks, but most occurred between the 4th and 10th. They were confined mainly to the ill and the elderly. The excessive death toll in this disaster is believed to reflect the synergistic effects of high levels of SO_2 coupled with high levels of smoke (particulates). The maximum daily SO_2 concentration was measured at the time as 1.34 ppm; the highest smoke level was 4.46 µg. m^{-3} (Perkins 1974). These are average values and much higher concentrations were doubtless experienced for short periods and in some areas.

The significance of this synergistic effect is illustrated in the fact that a subsequent 'killer smog' fell over London from 3 to 7 December 1962 (Scott 1963). Levels of SO_2 were similar to those of the 1952 event, but the 1956 Clean Air Act (see chapter 1) had by now brought a marked drop in smoke levels over London. The death toll in 1962 was accordingly lower. 340 died in the London Administrative County as opposed to 2,000 in 1952. Better awareness of the possible health hazard of prolonged smog, in the wake of the 1952 event (which received widespread publicity), is also believed to have been a factor in keeping deaths down in 1962. Many people with chest complaints had the sense not to go out of doors and expose themselves to risk.

These three disasters highlight the very serious effects of high levels of atmospheric SO_2 on human health. For every person that died, there were countless others who survived but suffered in various ways and to varying extents from irritation of the eyes, nose, and throat, coughing, difficulty in breathing, headaches, and vomiting. Apart from the toll on human life and the drain on human welfare, such pollution incidents have potentially massive economic implications – the costs of hospitalization and medical care, loss of productivity through absence from work, or inability to work at full strength, and so on.

Background levels and epidemiological studies

There are dangers in focusing *only* on disasters such as the three outlined above, because – by definition – they represent highly infrequent and unrepresentative extreme conditions. It is as relevant here to consider the effects of more 'normal' (that is, background, or ambient) levels of pollution on human health, which are generally evaluated in epidemiological studies (epidemiology is 'the study of the factors involved in the distribution and frequency of a disease process in a given population' –

Perkins 1974: 325). We must not overlook the great difficulty of establishing how individual pollutants affect human health because of the large
number of factors and variables involved (Durzel and Cetrulo 1981; Kim
1985; Lipfert 1985).

The medical literature on the epidemiology of SO_2 is wide and reviewed
elsewhere (Perkins 1974), but some generalizations can be drawn from it.
Most levels of SO_2 to which humans are commonly exposed are not fatal,
except to unusually sensitive individuals. But it is a major cause of illness,
disability, and discomfort.

Sulphur dioxide appears to affect the lower respiratory tract (i.e. lungs),
particularly in terms of infections, more than the upper tract (i.e. nose,
throat, and windpipe), which explains why it affects breathing. It also
explains why SO_2 promotes and aggravates disorders such as bronchitis
and emphysema. The effects on breathing may produce more than
discomfort; they bring reduced efficiency in undertaking any form of
physical activity (including walking, lifting, gardening, etc.). Athletes have
been found to perform much worse in polluted air than in clean air, but it is
not clear whether this is a direct causal effect or stems from discomfort
(e.g. from eye irritation). There were, for example, worries about the
possible effects on the health of athletes taking part in the 1984 Olympic
Games in Los Angeles, a city with a long history of smog problems (Elsom
1984), but these proved to be without foundation. The impact of air
pollution on body reactions is also illustrated in studies that have
established a statistically significant association between oxidant levels and
rates of motor vehicle accidents in Los Angeles (Perkins 1974), though
again the link may not be direct and causal.

Epidemiological studies also show that it is not just short-term exposure
to high concentrations of SO_2 that causes health problems. Long-term (at
least several years) exposure to low levels of SO_2 pollution (below
0.05 ppm) can produce detectable health damage to children and adults.
They also show that nitrogen oxides affect health. Laboratory tests have
shown that NO_2 (nitrogen dioxide) increases the susceptibility of animals
to bacterial infections, and such effects may also occur in humans. For
example, a study in Chattanooga, USA, found that high levels of NO_2
reduced the breathing efficiency of children and significantly increased
incidence rates of illnesses amongst parents and children (Shy *et al.* 1970).
A study of the north-east of England found that death rates from lung
cancer and bronchitis are up to twice as high in heavily polluted urban
areas as in the surrounding countryside, even taking into account smoking
habits. It concluded: 'the implication that air pollution might play an
important role in relation to health' . . . (especially lung cancer) . . . 'is too
strong to ignore' (*The Times*, 25 April 1978).

Studies like these indicate that long-term exposure to ambient levels of
air pollution by SO_2 and NO_2 *can* have serious and detectable adverse
direct effects on human health. The disaster studies and the epidemiological studies, viewed alongside circumstantial evidence, indicate that high

concentrations of SO_2 and NO_x (and doubtless other pollutants like ozone – Colbeck 1985) – i.e. dry deposition of acids – *can* cause direct harm to humans (most notably bronchial and lung complaints) via inhalation. Concentrations of SO_2 and NO_x high enough to cause serious health problems are normally found only close to emission sources – i.e. in major cities and industrial areas – and, even then, often only under unusual weather conditions (e.g. thermal inversion) and/or when emissions are unusually high (e.g. during winter, when large quantities of fossil fuels are being burned for heating and power generation).

The evidence suggests that SO_2 and NO_x levels in many major cities (such as London and New York) have been declining over the last decade as a result of air pollution control legislation. The implication is that disasters like those at Donora and London should not happen again, at least on such large scales, and that more general health damage from exposure to ambient levels might also be expected to decrease in the future.

Indirect health effects

There is as yet little evidence to link wet acid deposition with direct effects on human health. But this does *not* mean that wet deposition has no effect, or that its effects are insignificant. There are at least two important indirect routes by which acid rain might affect human health, and both involve toxic heavy metals.

Heavy metals (such as copper, zinc, cadmium, and mercury) are liberated when soils and deposits are acidified (see chapter 5). These mobilized contaminants are dissolved in soil water, so they can readily infiltrate groundwater or seep out or flow into rivers and lakes. They can also be taken in – along with moisture and nutrients – by plants (via root osmosis) and by the grazing animals that eat those plants. Thus groundwater that is drunk by humans and food (fish, meat, and vegetables) that is eaten by humans can be contaminated with toxins, which are taken into the body by digestion (Thornton and Webb 1979; Thornton and Plant 1980). The human body stores rather than releases this heavy metal material, and so high (and potentially very damaging) concentrations can build up by the accumulation of small doses over long time-periods. Cadmium and the other heavy metals are concentrated in plants, then animal bodies, and eventually in humans as they move up the food chain; those at the top of the 'pile' intake the largest (and most toxic) concentrations.

There are reports of edible fish being contaminated by toxic metals (mainly mercury) (Tolba 1983). This can result in mercury poisoning in humans; the Japanese have experience of this in the form of Minamata disease. Farming crops can also be contaminated by heavy metals (especially cadmium) through mobilization in acidified soils.

Drinking water supplies can be affected by increased levels of toxic heavy metals (particularly aluminium, manganese, copper, cadmium) in two ways (McDonald 1985) – from leaching from soils in catchments, and from corrosion of water storage and distribution systems (most notably from galvanized steel and copper water pipes). There is evidence of contaminated groundwater in southern Sweden, parts of Ontario in Canada, and in the Adirondack region in New York state (Tolba 1983; Anon 1984/5e). No reports have yet appeared of seriously contaminated water supplies in Britain.

Research is still in progress into the possible health consequences of drinking water with high levels of heavy metals, but there is evidence that it is associated with increased incidence of osteomalacia, an uncommon bone disease which in the past was linked with vitamin D deficiency and rickets (*The Times*, 14 September 1985). Research in the United States and Great Britain shows that many more cases of the disease are observed in areas with high concentrations of aluminium in water supplies (such as Manchester and Newcastle in England) than elsewhere. Contaminated groundwater has also been linked with health problems in babies. In some rural parts of west Sweden, babies have been suffering from diarrhoea on and off for months and this has only occurred in areas where drinking water comes from acidified wells and has an abnormally high copper content (La Bastille 1981: 675–6). Drinking water supplies in industrial countries are cleansed by proper treatment techniques. Problems are more serious where water supplies are taken from wells, springs, local ponds, or lakes (Ballance and Olson 1980). For example, in Sweden over 1 million people rely on water drawn from their own wells, and about half of them live in areas where acidified lakes are common (Swedish Ministry of Agriculture 1984). High concentrations of heavy metals have been recorded in acid well waters in Sweden.

The acidification of groundwater brings other problems, too. Amongst the most colourful is the effect of people in southern Sweden washing their blond hair in water from copper pipes that have been corroded by acidification – the copper sulphate from the plumbing discolours the tap water, but it also turns the hair bright green (La Bastille 1981: 659)! The only options for affected families are to replace the copper plumbing and install filters, dig deeper wells, or learn to like green hair.

CONCLUSION

It is clear that acid deposition *can* cause serious damage to buildings and materials, and it *can* bring serious health problems for humans. Both wet and dry deposition create problems, in both direct and indirect ways. Most of the damage to buildings and humans arises in urban and industrial centres (rather than across the countryside at large), reflecting the dominance of local sources of emission of SO_2 and NO_x and the dominance of short-range dry deposition. The costs of damage to buildings and humans are far from insignificant, and the inconvenience and hardship so created are far from acceptable.

Over two decades ago the following conclusion was drawn about the problems of air pollution from smoke:

> In towns and industrial districts rain water loses its purity; ash and other solids fall continuously to the ground; the air contains a suspension of fine particles which penetrate indoors, to be deposited on walls, ceilings, curtains and furniture; our clothing, our skins and our lungs are contaminated; metals corrode, buildings decay, and textiles wear out; vegetation is stunted and blackened; sunlight is lost, germs multiply; our natural resistance to disease is lowered. In a hundred and one ways the miasma of atmospheric pollution is lowering our vitality and our enjoyment of life.   (Meetham 1964: 4)

Since the mid-1950s in Great Britain and the late 1960s in the United States legislation has attempted to reduce this component of air pollution, and it has been successful in doing so. However, the legislation has *not* been entirely successful in reducing air pollution, and a similar conclusion can be drawn today about the problems of pollution from acid gases, mists, and aerosols.

Part III

THE TECHNOLOGY
OF ACID RAIN

7 CURES AND REMEDIES
'We have the technology . . .'

Progress, therefore, is not an accident, but a necessity . . . It is a part of nature.

(Herbert Spencer, *Social Statics*, 1850, pt i, ch. 2: 4)

We have seen in Part II that considerable scientific uncertainty still surrounds many aspects of the effects of acid rain, despite much research in many areas. Some critical questions remain as yet largely unanswered, such as the linearity of the relation between oxide emissions and acid deposition, and of that between levels of deposition and observed damage. As a result there is still *no* universally accepted consensus on the need for remedial measures, or on what types of measure might be most appropriate.

In this chapter we shall examine the main proposed solutions to the problems of acidification in the environment. They fall into two groups – cure and prevention. Curative measures can be applied where the problems (e.g. acidification of soils and surface waters) actually arise. Preventive measures can be applied at source (i.e. at points of emission of the sulphur and nitrogen oxides). The latter are more effective, more sustainable, and more urgently required. They are also the more expensive (at least in the short term) and the least acceptable to (and frequently overlooked by) industry and to some governments.

CURE BY BUFFERING

It is common sense that problems can best be dealt with by making sure that they never happen in the first place; but if they *do* (or are *allowed* to) happen, then they must be solved quickly and economically (i.e. cost-effectively). Ideal solutions are also sustainable and cause minimum interference. The only acceptable long-term 'solution' to problems of acidification is prevention, but in the short term there is a pressing need to cure the damage it has already caused. This requires not simply the removal of symptoms of damage (e.g. restocking fish in acidified lakes and rivers, planting new trees in areas affected by forest dieback, or cleaning discoloured building facades); it also requires restoration of natural chemical balances to ensure that damage does not recur.

We have already encountered the natural ability of some materials to buffer, or effectively neutralize, or counteract acid inputs. Lime and limestone are the most popular of a range of chemicals (including caustic soda, sodium carbonate, and dolomite – Tolba 1983: 118) that can be used to buffer acidic materials. Limestone bedrock and lime-rich deposits offer a natural buffer capacity to soils and catchments (chapter 3); similar chemical reactions explain why limestone and dolomite building materials can be badly attacked by dry and wet acid deposition (chapter 6).

This buffer capacity of lime-based materials has been eagerly seized upon as a basis for 'curing' (i.e. counteracting) some of the biological damage caused by acidification. Lime has been added – normally by aerial spraying from helicopters – to rivers, lakes, soils, and forests in a number of areas to reduce acidity, alleviate damage, and improve conditions for plants and animals (Anon 1984/5c; McCormick 1985). The approach has been most widely explored in Sweden, where liming of acidified lakes began on an experimental basis in autumn 1976. Within two years, some 200 lakes had been limed, and a further 700 were limed between 1977 and 1979 mostly with government support (Bengtsson, Dickson, and Nyberg 1980). By summer 1985 over 1,500 lakes had been limed (Swedish Ministry of Agriculture 1984). Liming raises the pH of water, thus improving conditions for plants and animals. It fixes the dangerous aluminium ions (see chapter 4), which are then precipitated on to the lake bed, and it also reduces the freely available mercury in the water – both to the advantage of fish and other aquatic species.

Sweden's experience is one of escalating costs: the budget of $2.5 million for 1980 (Bengtsson, Dickson, and Nyberg 1980) had risen to $7.5 million by 1983 (May 1982: 11), and it is expected to rise much further. This is clearly an expensive 'cure', but it has produced encouraging results (e.g. recolonization by indigenous phytoplankton, zooplankton, and fish – Bengtsson, Dickson, and Nyberg 1980) and is regarded as the only practical alternative *without* or *pending* widespread adoption of emission-reducing technology.

Liming has also been used as a means of restoring acidified soils, at least temporarily. The calcium counteracts the harmful effects of acidity and aluminium, and improves the productivity of acidified croplands and forests. This form of land management is far from new, because lime has for centuries been added to acidic and calcium-poor soils as a means of improving texture and chemical status (although the application of nutrient fertilizers usually has the reverse effect of increasing soil acidity – Friends of the Earth 1985).

What is new, and somewhat unacceptable, is the need to use liming to counteract the disturbing influence of acid deposition derived from air pollution. This spreads the true cost of air pollution amongst farmers (see, for example, Figure 5.6) and foresters, rather than being met by those who produce the pollutants (the power stations and industrial point sources). It also speeds the long-term deterioration of soil productivity, because lime cannot simply be added to a given soil in growing quantities, seemingly for ever, without adverse wholesale changes in soil status.

Also unacceptable is any growing reliance on the need to add lime to soils simply to maintain farming productivity in areas that receive large inputs of acids from air pollution. In Britain, the need for liming in marginal upland farming areas was for many years formally recognized by the provision of grant support from the Ministry of Agriculture, but this grant support was removed in 1976. Since then many farmers have either ceased liming, or apply lime less frequently; total applications of lime on British farms fell from 4.6 million tonnes in 1976 to 3.2 million tonnes in 1984 (Friends of the Earth 1985). There are fears that this reduced buffering might speed soil acidification, reduce soil fertility, and possibly ultimately lead to declining agricultural yields (hence increase Britain's dependence on food imports).

Liming does alleviate some of the worst symptoms of acidification, but it is *not* a real or sustainable cure because it does not attack the root causes of the problems (Tolba 1983: 118; Hjelm 1984). As a means of buying time it has wide potential, but there is no escaping its inherent drawbacks. The benefits can be short lived, in that it works for a time then the effects become neutralized by further acid inputs (the dosages added to Swedish lakes are designed to last for five years – Bengtsson, Dickson, and Nyberg 1980). Moreover, it is not practicable or suitable for many lakes and rivers; sometimes liming is of *no* help at all. Liming also produces its own problems, through the deposition of the toxic heavy metals (e.g. aluminium) previously mobilized in the lake waters, which poison lake bottom dwelling organisms (Swedish Ministry of Agriculture 1984). It is certainly a costly solution, and one that needs repeating frequently.

Despite the conclusion by Britain's Watt Committee (1984) that the liming of exposed lakes may be the most effective means of control for acidified lakes, which may cost only 'a few percent of measures to retrofit power stations', liming is an interim measure that provides a short-term biological defence. It is 'a sort of artificial respiration for dead lakes and streams' (May 1982: 11), a means 'of keeping poorly buffered or acidified waters alive' (Bengtsson, Dickson, and Nyberg 1980: 35) – until oxide emissions can be reduced to adequately low levels. This can *only* happen through a deliberate and sustained policy of prevention, rather than cure.

PREVENTION

The focal point of the political debate over acid rain (see Part IV) is the need to reduce rainfall acidity by controlling emissions of SO_2 and NO_x at source; i.e. *prevention*. Across the United Kingdom as a whole about a quarter of the acidity is derived from NO_x, the other three-quarters from SO_2 (Clarke 1985: 5). About two-thirds of the SO_2 and nearly half of the NO_x comes from the burning of fossil fuels (mainly coal and oil) in conventional power stations (Figure 7.1).

Figure 7.1 Main sources of emissions of sulphur dioxide and nitrogen oxides in the United Kingdom, 1980

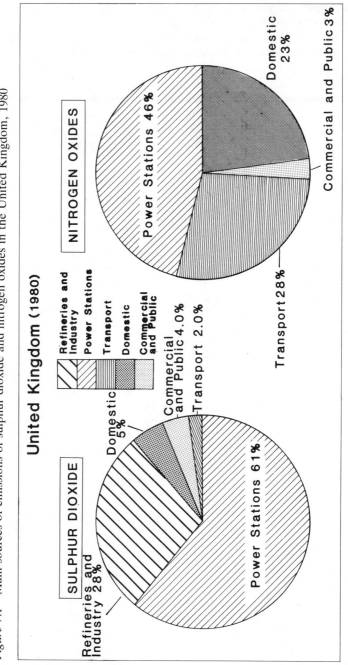

Source: based on data in Table 2.3, after Scriven (1985)

It is understandable, in the light of this, that so much attention in the debate has centred on power stations as major villains in the piece, and that so many and well-orchestrated calls have been made to ensure that power stations cut their oxide emissions sharply and without further delay. It is therefore of more than passing interest to consider what means exist to reduce emissions of SO_2 from power stations, and emissions of NO_x from power stations and vehicles.

Reducing emissions of SO_2 from power stations

All fossil fuels contain some sulphur, and coal has a higher sulphur content than oil and gas. Sulphur occurs naturally in coal in three forms – as iron pyrites (FeS_2, an inorganic sulphide), organic compounds (bound into organic matter), and sulphates. Only the first two are of major importance by quantity or impact (Perkins 1974: 263). However, the total sulphur content of coal varies considerably from deposit to deposit: in commercial coal it normally ranges between 0.3 per cent and 5 per cent (by weight). A sulphur content above about 1 per cent – typical of the North American coal mined from the vast deposits east of the Mississippi (Priest 1973: 75) – is regarded as 'high'.

As described in chapter 2, oxides are formed when a fossil fuel is burned at high temperature in air. The amount of oxides produced depends in part on the sulphur content of the fuel source (both increase together) and in part on the temperature of combustion (again, both increase together). When coal is burned, the main oxide produced is sulphur dioxide (SO_2), the key ingredient in the acid rain debate.

There is only one way of eliminating SO_2 emissions altogether, and that is to stop burning fossil fuels for energy production. Given our heavy reliance on available energy today, such a solution is *not* viewed as a practical option. The alternative is to reduce SO_2 emissions, and the serious search for viable methods of doing this has been on for over a decade (Ludwig 1969: 23). For example, in 1973 ten different schemes were being tested on large (> 100 MW) power stations in the United States alone, and a further twenty were at the prototype stage (Perkins 1974).

Some control technologies have emerged as more economically viable, energy efficient, or operationally practical than others. What follows is an attempt to summarize in a non-technical manner the main options presently available; technical treatments are available elsewhere (Perkins 1974; Lynn 1976; Elsworth 1984a).

Option 1. Burn less fossil fuel

This is seen by many scientists and all conservationists as the best and certainly the most permanent solution to the problem of reducing SO_2 (and NO_x) emissions at source. The goal is to promote long-term reductions in consumption of coal and oil in conventional power stations. This can be achieved in two ways, applied separately but preferably together.

(a) *Energy conservation*: Reduced fuel consumption will follow from a successful and sustained energy conservation programme. A significant reduction in overall energy use in developed countries in the future seems highly unlikely, but there are some prospects of stabilizing and real prospects of reducing the rate of growth of energy use. This would require more efficient use of existing fuels, and technical improvements could ensure that processes burned fuel more efficiently (Tolba 1983: 118). It would also require improved thermal insulation of buildings.

It would certainly be both possible and sensible to pass on to the energy user some degree of responsibility for more careful use of available energy resources. For the industrial user this might include tax concessions or other financial inducements for installing energy-efficient technologies. For the individual domestic energy user, the choice ranges from the simple act of switching off lighting and heating wherever possible, to the more demanding act of adjusting purchasing strategies in favour of energy-efficient goods. Most individual energy users could reduce their energy consumption without facing undue personal hardship or inconvenience – indeed they would reap the benefit of lower energy bills, as well as the satisfaction of knowing that *they* have played a role (no matter how small) in reducing acid rain!

Electricity (externally supplied) is widely viewed by the individual user in the same way that they view water from the tap (externally supplied) – freely available in abundance to be used where, when, and how you decide, and at worst you pay for what you use! Consumers need to be made aware of the folly of this view, and this is best done by 'awareness building' campaigns (rather than through cost levels designed to promote less profligate use). Government initiatives that educate energy users – like Britain's national 'MONERGY' campaign of 1986 – serve a key role in this respect.

All such conservation schemes are long term, so that any contribution they might make to overall reductions in oxide emissions will be extremely difficult to detect and quantify. But the post-1974 'energy crisis' – reflected in wildly escalating energy costs and brought about largely by OPEC oil price rises – has encouraged more considered use of energy and the search for suitable alternatives to fossil fuels.

(b) *Substitution of fuel sources*: Oxide emissions from power stations can also be reduced by switching at least a portion of the total energy production capacity to non-fossil fuel sources. Substitution is also seen as a long-term solution, with sulphur emissions declining slowly over the period of replacement.

There are a number of sulphur-free substitutes to choose from. Energy can be derived from wood, natural gas (methane), petrol (gasoline), paraffin (kerosene), and gas oil. Electricity can also be generated from renewable energy sources such as the sun, winds, tides, waves, falling water (hydro-electricity), and natural heat in the earth's crust (geothermal energy). There are many research initiatives under way in so-called 'low energy' (or 'alternative' or 'soft') technology, which are based on use of renewable sources of energy (Sawyer 1986). But most such sources are limited by geographical factors and they require vast energy storage capacity to match supply and demand in both time and space.

Each of these substitutes might offer some scope for local use, particularly in isolated communities, but their potential for supplying anything like the national requirements of energy in countries like the United Kingdom and the United States is very limited by geographical, practical, and economic factors. The only viable substitute, given today's scientific understanding, is nuclear power (Lynn 1976) which – even its sternest critics (like Friends of the Earth 1985) concede – has vast potential.

Conventional nuclear power stations generate electricity from radioactive uranium using the fission process, which involves converting small amounts of uranium's mass directly into energy (Lynn 1976). Uranium reserves world-wide are somewhat limited, however, and so recent interest has centred on the scope for using plutonium in the fission process. This requires 'breeder' (or 'fast-breeder') reactors, which 'breed' fuel by converting non-fissionable (i.e. unusable) types or isotopes of uranium into fissionable (i.e. usable) fuel. Research is under way into the next generation of reactors, which will tap the nuclear fusion process (in which two small atoms of the deuterium isotope of hydrogen are fused together into a larger helium atom, which contrasts with the fission process in which large uranium atoms are split into smaller pieces), but large, commercially viable, and operationally safe fusion power stations are – as yet – some way into the future.

Nuclear power presently supplies less than 5 per cent of the energy used in the United States (Sawyer 1986) and around 8 per cent of Britain's energy (Fishlock and Wilkinson 1986), but both countries have a commitment to expand the nuclear power base (and ongoing programmes to do so). In one sense, expansion of nuclear power would make a useful contribution to reducing acid rain, because nuclear plants release no gas pollutants containing sulphur or nitrate compounds. But to argue in favour of a nuclear programme simply from this perspective would be completely to overlook the immense problems of converting and storing the radioactive waste products so unavoidably produced in nuclear power stations. It would also be to overlook the very real dangers of nuclear reactors, and the health hazards they pose to humans – which are both direct (contamination with radioactive fallout) and indirect (via contaminated food and water supplies). In the wake of the Chernobyl (USSR) disaster of March 1986 (Serrill 1986), and the 'near-miss' at Three Mile Island, Harrisburg (USA) in March 1979 (Bunyard 1979), such an oversight would be hard to defend!

We must conclude that the option of large-scale substitution of fuel sources is not without its inherent limitations and attendant dangers, and concede that Option 1 might make a contribution to the oxide emission problem, but not the largest one.

Option 2. Switch to low-sulphur fuel

We noted above the natural variability in sulphur content of fossil fuels. Given our heavy reliance on conventional means of producing energy – which in practice means burning coal, oil, and gas – it is logical to seek to minimize oxide emissions at source by burning fuel with the lowest possible sulphur content. This option offers one of the easiest ways of controlling SO_2 emissions from power stations, which does not require costly investment in pollution-control technology (and high recurrent running costs).

In the United States during the late 1960s, there was mounting public pressure on national, state, and local governments to curb air pollution, and this was the only practical means of limiting SO_2 emissions then available. Consequently regulations limiting the sulphur content of fuels for use in combustion installations (power plant, industrial burners, etc.) were passed in a number of cities and states (Lynn 1976). This approach was adopted in New York City in 1965, when an upper limit of 1 per cent sulphur in fuel was imposed. A significant drop in SO_2 levels over the city, especially at low (< 0.5 ppm) concentrations, was noted over the following years (Eisenbud 1970). Washington DC also forced the General Services Administration to switch to lower-sulphur fuels in their heating installations, and an improvement in air quality followed (Lynn 1976).

An interesting variant on the 'low-sulphur fuel' option has been tried in the United States, where some large power stations have been designed with the capacity to switch fuel sources as and when required. For example, some stations routinely switch fuel from season to season – partly to capitalize on variable fuel costs and partly to minimize sulphur emissions under particular weather conditions (Lynn 1976). Some power stations in urban areas switch to low-sulphur fuel for short periods (Lynn 1976), when adverse weather conditions would cause serious pollution episodes (see chapter 6).

This option does, therefore, offer some prospect of contributing to a net reduction in SO_2 emissions. But it does have its limitations. A critical constraint is the limited availability of low-sulphur coal in most countries. In the United Kingdom, for example, only a small fraction (15 per cent) of the available coal has a sulphur content less than 1 per cent (Table 7.1). Potential supplies of low-sulphur coal in the United States are much higher, but most reserves are long distances from power stations so that transport costs are very (sometimes prohibitively) high (Priest 1973: 75).

An added constraint, closely related to relative shortage, is the relatively high cost of low-sulphur fuel. North American experience during the 1970s has shown how demand-led competition – many users switched from high-

Table 7.1 *Sulphur content of United Kingdom coal, 1982–3*

Source area	Average sulphur content (%)	Saleable output (million tonnes)
Scottish	0.70	6.6
South-west opencast	0.90	2.1
South Wales	0.95	6.9
Scottish opencast	0.95	2.8
South Notts	1.37	8.3
North Notts	1.48	12.4
Doncaster	1.48	6.8
North-east	1.49	12.4
Western	1.52	10.8
South Midlands	1.56	8.2
North-east opencast	1.56	3.1
South Yorks	1.59	7.3
North Derbyshire	1.69	8.1
Barnsley	1.85	8.1
North Yorks	1.92	8.4
Central east opencast	1.96	3.2
Central west opencast	2.07	2.5
North-west opencast	2.34	1.0
Total	1.51	119.0

Source: Dudley, Barrett, and Baldock (1985), Table 6

to low-sulphur fuels to comply with air-quality regulations (many from coal to oil) – forced a marked rise in the price of available low-sulphur coal and oil (Lynn 1976). Any further widespread switching to low-sulphur fuels will force further price rises that could well make this a non-viable option.

There is also a technical limitation in using low-sulphur fuel in conventional power stations, because such fuel decreases the efficiency of precipitators, which are used to remove fly ash and particles from exhaust gases (Priest 1973). There is thus the risk that increased reliance on low-sulphur fuels might reduce sulphur gas emissions but increase particulate pollution; the trade-off is of questionable value.

Both options *would* reduce SO_2 emissions from power stations, but most observers agree that emissions in Europe and North America will only be reduced to an acceptable level if more concerted efforts are made to remove the sulphur from fossil fuels and from emission gases. There are three main alternatives to choose from (Persson 1976: 249): reduce the sulphur content of fuel before combustion, reduce the production of SO_2 during combustion, or reduce emissions of SO_2 after combustion but before exhaust gases are emitted from chimneys into the atmosphere.

Option 3.   Fuel desulphurization

Techniques are now available to reduce the sulphur content of fossil fuels
before they are burned, so that a given amount of desulphurized fuel will
produce much less SO_2 on combustion than would the same amount of
uncleaned fuel. The sulphur content of natural gas is routinely removed at
source (Lynn 1976) and both oil and coal can also be 'cleaned' (the latter is
a bit less viable). The two main methods involve either washing or
chemical removal of the sulphur.

Coal washing has long been undertaken as a routine preparatory procedure
to remove ash from freshly mined coal before it is transported to power
stations to be burned. But the washing also removes some of the pyritic
sulphur from the coal (Clarke 1985: 5) – a significant gain at marginal cost –
so this process has been used in Europe and North America to clean the
coal before combustion. Washing has been more extensive in North
America because of the higher sulphur content of its indigenous coal. The
economics of each colliery normally determine how much coal is washed
on site (Meetham *et al.* 1981: 201).

The process is simple and normally takes place at the coal preparation
plant. Coal is crushed and ground into tiny (dust-sized) particles, then it is
floated on water. The finely ground coal particles float, but the heavier
minerals sink. The washed coal is then dried in hot air; normally this is
simply a stream of hot exhaust gases derived from coal burning (around 2
per cent of the total coal stock is used for drying – Persson 1976: 249). The
process is very efficient.

Washing removes up to 30 per cent of the raw coal as ash rejects
(Persson 1976: 249), and normally 40–60 per cent (by weight) of its pyritic
sulphur content (Tolba 1983: 118). But this often reduces the SO_2 gas
emissions from the coal by as little as 8 per cent (Persson 1976: 249) to 15
per cent (Meetham *et al.* 1981: 201), because washing removes only some
of the inorganic sulphur, and it cannot remove any of the organic sulphur
that is chemically bound to the coal (Priest 1973: 75).

The cost of washing crushed coal is relatively cheap – quoted values are
in the range $1–6 per tonne of coal (Tolba 1983: 118) – but there remains
some doubt about the cost-effectiveness of such treatment (Priest
1973: 75). Such doubts are amplified if the accounting is based solely on
operating costs, and values are *not* attached to having clean air.

Chemical cleaning methods for coal are as yet still being developed (Tolba
1983: 118), but they have the real advantage of removing both the organic
and the inorganic sulphur from the coal (Persson 1976: 249). They are also
more expensive than washing – up to 95 per cent of the pyrites (inorganic)
and half of the organic sulphur can be removed at a cost of $20–30 per
tonne of coal (Persson 1976: 249). Washing adds 1–6 per cent to electricity
costs; chemical desulphurization adds 15–25 per cent (Tolba 1983: 118).

Sulphur-rich coal can also be converted into a burnable gas or a liquid, with the sulphur removed in the process (Priest 1973: 73). These processes are referred to respectively as *gasification* and *liquefaction*, and both can significantly reduce the sulphur content of the fuels. In the United States, where natural gas reserves are very limited, large quantities of high-sulphur coal have been converted to gas with the double advantage of reducing SO_2 emissions and gaining more productive use of available energy reserves (Lynn 1976). Gasified coal is a relatively expensive fuel source (Meetham *et al.* 1981: 301).

Liquid fuels can also be desulphurized; recent chemical advances make it possible to remove up to 99 per cent of the sulphur from most crude oils (Meetham *et al.* 1981: 301). Desulphurization of gas oil is now widely done in new refineries in Europe (Persson *et al.* 1976). Up to 80 per cent of the sulphur can also be removed from residual oil, but economic factors (it increases costs by 20–35 per cent – Lynn 1976) dictate how widely this is in fact done. Because gasoline oil is so profitable, the United States produces relatively little residual oil; but many overseas refineries produce more and cheaper residual oil (some with a high sulphur content) (Lynn 1976).

The costs of desulphurization of liquid fuels appear to be prohibitively high. Estimates given to the 1982 Stockholm Acidification Conference (see chapter 8) suggest that it costs $20–40 per tonne of fuel oil to remove sulphur, which would add 10–20 per cent to the cost of producing energy (Tolba 1983: 118).

On balance, therefore, fuel desulphurization is technically possible but economically questionable as a basis for reducing sulphur emissions at source.

Option 4. Sulphur reduction at combustion

The best alternative to using low-sulphur fuel is to use fuel with a higher sulphur content but ensure that most of the sulphur is removed during combustion – before it ever has the prospect of becoming an exhaust gas (SO_2) emitted from chimneys. This can be done using *fluidized bed technology* (FBT), which fixes sulphur to lime at the time of combustion and 'offers a promising way of reducing pollution by both sulphur and nitrogen oxides in heat and power plants' (Tolba 1983: 118).

FBT is a very efficient means of modifying combustion to reduce SO_2 production (Persson *et al.* 1976; Allar 1984; Clarke 1985), but it requires a special type of boiler (the 'fluidized bed boiler') in which the coal is burned. The process exploits the known buffering ability of limestone. Coal is fed into the combustion chamber from one side, and finely crushed limestone (or a similar material capable of absorbing SO_2 at combustion temperature) is fed in from the other. Thus the coal is burned in the presence of limestone. This takes place on a grate through which air or gas is passed from below; the turbulence causes constant agitation and thus complete mixing of the coal and limestone (hence 'fluidized bed'). This mixing increases the efficiency of the combustion process (*all* of the coal is burned properly), but it also ensures that all of the coal is exposed to the buffering agent during combustion. The buffer effectively strips the coal of its sulphur, forming an ash residue (a mixture of calcium sulphate, calcium oxide, and calcium sulphide). Tests show that the fluidized bed technology is capable of removing nearly all of the sulphur in the fuel (Persson *et al.* 1976).

There are a number of economic advantages in the FBT option. The technology is now commercially available to small and medium-sized plants, at costs comparable to conventional combustion systems (Swedish Ministry of Agriculture 1984). It involves no great additional capital expenses, thus is quite suitable for any new power stations that might be planned (it is not an 'add-on' or retrofit technology – like flue gas desulphurization (option 5) – which requires costly conversion of existing plant). Moreover, the sulphur enriched limestone ash is a saleable product, and can be used commercially as gypsum (Persson *et al.* 1976). But there are other advantages than simply the economic ones. Because the fuel is burned at a much lower temperature than in conventional boilers, emissions of NO_x are also reduced considerably (Persson *et al.* 1976) (rate of production of NO_x depends in part on combustion temperature).

There is a growing belief that FBT is the most promising prospect for reducing SO_2 from high-sulphur fuels, but the newness of the technology cannot be overlooked. In some senses it remains (in late 1987) a *prospect*, rather than a *reality*, because although scale trials have proved successful there is still some uncertainty over how costly, cost-effective, or operationally reliable the technology will turn out to be. The technology is believed to be feasible and economic in coal-fired power stations capable of producing up to 250 megawatts of energy (Swedish Ministry of Agriculture 1984).

Experimental work on fluidized beds, carried out on behalf of Britain, West Germany, and the United States, began in Britain in the early 1970s. From 1980 onwards testing has centred on a large trial furnace at Grimethorpe Colliery, near Barnsley (Yorkshire) (*The Times*, 25 May 1984, p. 14), and further evaluation tests by the Central Electricity Generating Board and the National Coal Board on larger boilers began in 1985. Pending the outcome of such tests, the long-term and widespread applicability of FBT is difficult to judge.

Option 5. Flue gas desulphurization

One alternative to removing sulphur from coal during combustion is to remove the sulphur gases produced in combustion before they are released into the air (i.e. emitted from power station chimneys). This involves 'effluent cleaning', or – more correctly – flue (or stack) gas desulphurization (FGD), which is a relatively recent development in commercially viable terms and still very much a new area of research (Maurin and Jonakin 1970; Slack 1973).

It is much more difficult to remove sulphur after combustion than before or during because it is dispersed (in exhaust gases) rather than concentrated (in coal). Sulphur gas is about a thousandth of the density of sulphur in solid form (Meetham *et al.* 1981: 201), so that large volumes of exhaust gas with small concentrations of SO_2 in it have to be treated to remove equivalent amounts of the pollutant. FGD removes sulphur oxides from flue gases by chemical reaction that encourage absorption of the sulphur on to a solid or liquid. There are three main techniques to choose from.

The *dry regenerative* approach involves passing the flue gas through a bed of dry absorbant (such as activated carbon). The absorbant reacts with the SO_2, and by capturing and retaining the gas (on the surfaces of the particles) it prevents SO_2 emissions in the flue gas. The SO_2 can be chemically removed from the absorbant, then converted into sulphur (S) or sulphuric acid (H_2SO_4) to be sold (Persson *et al.* 1976). The process is 'regenerative' in the sense that it produces usable sulphur as a by-product, the sale of which helps to offset the cost of removal (Priest 1973: 75).

Wet techniques have been used more extensively, and are likely to be adopted more widely in the future. In the *wet regenerative* approach the flue gas is passed through a liquid (the reagents are normally dissolved in water) in which the SO_2 is absorbed. The sulphur, now in solution, can then be converted to yield saleable S or H_2SO_4 (Persson *et al.* 1976), again helping to offset costs of removal. This technique is commonly referred to as 'wet scrubbing', and it normally involves bubbling the dirty flue gas up a packed tower in which the contaminants are absorbed; clean (i.e. desulphurized) gas leaves the top of the tower (Onnen 1972).

The *wet non-regenerative* approach involves passing the flue gas through a 'scrubber' containing a water slurry of lime or limestone. The technique is also referred to as 'limestone scrubbing' (Persson *et al.* 1976). The SO_2 and sulphate solids in the flue gas are removed by direct buffering on to the surface of the lime or limestone particles, producing calcium sulphate (gypsum). This approach is non-regenerative because the absorbant (i.e. the slurry) is thrown away after use and its sulphur content is not reclaimed and reused.

Limestone scrubbing is the most common approach to effluent cleaning at present (Tolba 1983: 118; McCormick 1985). It also has a long history. London's Battersea power station (built in 1929) was the first installation in the world to have a successfully operating FGD system, which cleaned 80–90 per cent of the sulphur from the flue gas for most of its working life (Elsworth 1984a: 61). Scrubbing equipment was also installed at Fulham power station (built in 1935) and at the oil-burning Bankside power station (built in 1954), both in London (Meetham *et al.* 1981).

Scrubbing has some distinct advantages. It can be an 'add-on unit' for existing power stations, so it does not require the building of new stations. In fact in the United Kingdom all large modern power stations are now required, at the planning stage, to provide enough space on site for FGD to be installed (retrofit) if necessary (Elsworth 1984a: 61). In addition, FGD does not require the electrostatic precipitators fitted to conventional power stations to reduce particulate emissions in flue gas (Oglesby 1971; Perkins 1974: 263). Moreover, it removes particulates and some NO_x from flue gas, as well as the SO_2.

But – as ever – these benefits are gained at a cost! Sludge disposal is a real problem. Large quantities of wet sludge with no commercial use or value are produced (around half a tonne of sludge – 60 per cent solid, 40 per cent liquid – is produced per tonne of coal; Persson *et al.* 1976), which generally has to be dumped. Another drawback is that calcium is deposited on the equipment, which reduces its operating efficiency and requires regular cleaning (at a cost) (Lynn 1976). The wet scrubbing process also reduces the temperature of the flue gas, so the gas must be reheated (at a cost) after scrubbing to give it adequate buoyancy to ensure that it does not fall to the ground close to the power station (Persson *et al.* 1976) (see Figure 2.7).

Wet scrubbing normally increases electricity costs by 8–18 per cent (Highton and Chadwick 1982) and removes up to 70–90 per cent of the SO_2 from flue gas (Clarke 1985: 5). Despite its proven abilities, many large sources in the United States have been extremely reluctant to use FGD processes to reduce sulphur emissions, arguing that their feasibility is *not* yet fully established (Lynn 1976). They favour continued reliance on burning low-sulphur fuels (option 2) without flue gas cleaning. This has increased demand for remaining low-sulphur fuel, pushed its price up, and made the long-term suitability of option 2 somewhat questionable.

Option 6. Disperse flue gases

There is an alternative strategy to reducing SO_2 emissions from point sources, and that is to disperse the oxides over as wide an area as possible. This can be achieved by releasing flue gases into a region in the atmosphere where they will be relatively harmless, either by using tall enough stacks or by building the power stations in remote places (Meetham *et al.* 1981).

This was the thinking embodied in the 'tall stacks policies' favoured in Britain and the United States some twenty years ago. But – as we have seen in chapter 1 – the approach is now regarded as unacceptable because, whilst it *did* serve to disperse gaseous emissions over a wider area, that very process has played a key role in producing the acid rain that is now so widespread a problem (see Figures 2.6 and 2.7). It has turned what was previously a local problem of dry deposition into a regional and trans-frontier problem of wet deposition. It is difficult *not* to conclude, with the benefit of hindsight, that this option contributes to rather than helps to solve the problems of acid rain.

Reducing emissions of NO_x from power stations

Most attention in the acid rain debate (see Part IV) has rightly centred on reducing emissions of SO_2 from power stations. Whilst the contribution of power stations to NO_x emissions (producing around 46 per cent of the man-made NO_x in the United Kingdom – Figure 7.1) is smaller in terms of quantities produced (see Table 2.2) and arguably less significant in terms of biological effects (chapters 4 and 5) and damage to buildings and humans (chapter 6), it must *not* be overlooked.

Nitrogen is not present in the fossil fuels burned in conventional power stations; it comes from the air introduced during combustion. Consequently, control technologies for the reduction of NO_x must be applied during and after burning (Clarke 1985: 5); switching fuel sources or cleaning fuel before burning have minimal impact on NO_x emissions. Most available techniques are based on relatively simple technologies and do *not* require large capital costs. There are five main options (Perkins 1974: 263), four of which reduce NO_xemissions *during* burning.

Because the nitrogen comes from air introduced during combustion, one approach (*low excess air combustion*) is to minimize the amount of air present at the point of combustion. In most burners, there is normally 10–20 per cent of excess air (i.e. air additional to that required *just* to complete the combustion process). If this excess air can be reduced to a minimum (ideally zero), then NO_x production will be reduced significantly. Such reductions can be ensured on oil- and gas-fired boilers, and are also believed to be possible on coal-fired boilers.

The rate of production of NO_x also depends on the combustion temperature in the boiler. Thus an alternative approach involves reducing the peak temperature within the boiler. In *two-stage combustion*, about 95 per cent of the air required for combustion is released in the boiler *at* the burner, and the rest is injected *above* the burner (to complete combustion at stage two). This reduces the peak flame temperature, and can reduce NO_x production in gas- and oil-fired boilers by up to 35 per cent. A variant is to *modify burner design* to allow for tangential firing; here the burners are located in the corners of the boiler, and they fire at a tangent to a circle in the centre of the furnace. Because the flames in the corner-fired boiler interact with one another less than in a conventional boiler, combustion is more efficient and it can take place at a lower peak temperature. NO_x production is reduced accordingly, sometimes by up to 30–40 per cent (Clarke 1985: 5).

In *flue gas recirculation* a portion of the exhaust gas is piped back to the combustion chamber. This flue gas has a very low oxygen content, so its presence in the chamber reduces peak temperature. NO_x production can be reduced by up to 90 per cent by simple recirculation of the flue gas.

The final option involves reducing NO_x levels *after* burning, by *flue gas cleaning*. It is much more difficult to remove NO_x than SO_2 from flue gas, partly because NO_x are very stable (thus they do not react very readily with absorbants) and partly because they are present in much smaller concentrations (about one-third of the SO_2 levels). As noted above (SO_2 option 5), some of the NO_x component of flue gas can be removed in limestone scrubbing. But the technique is of limited effectiveness – normally only about 20 per cent can be removed (Perkins 1974).

The control techniques for removing NO_x from power stations and industrial boilers are thus of variable effectiveness. Those that involve modifications of the combustion process seem to hold out most promise for the future (Perkins 1974).

Reducing emissions of NO_x from vehicles

Although it is difficult to arrive at precise figures, it is estimated that just over half (51 per cent) of the man-made NO_x in the United States (Lynn 1976: 63) and around 28 per cent of the man-made NO_x in the United Kingdom (Figure 7.1) comes from transport, being emitted in the exhaust gases of vehicles (mostly from petroleum engines). There is a much larger number of non-stationary sources of NO_x than stationary sources of SO_2, each one contributing a tiny fraction of the overall pollution budget. NO_x sources are therefore much more widely dispersed and more mobile; NO_x production is more variable through time and from place to place.

There are a number of strategies for reducing NO_x emissions from vehicle exhausts. Here is not the place to explore in great detail the technology of vehicle engines (Lynn 1976: 238–43), but we should note the main pollution control alternatives. Vehicle manufacturers in the United States have, since the late 1960s, wrestled with the problem of meeting increasingly stringent federal pollution-control standards. Consequently the United States enjoys a monopoly on initiatives and practical experience in this field.

Option 1. Modify engines or exhausts to reduce emissions

The two most common approaches (Lynn 1976; Strauss and Mainwaring 1984; McCormick 1985) are either to remove or alter the pollutants within the exhaust system, by installing appropriate technical devices, or to alter the operation or design of the engine to minimize production of the pollutants in the first place. Much of the early research (especially in the United States during the 1960s) was directed towards the former, and involved mufflers and after-burn systems. These proved *not* to be entirely successful in reducing emissions of NO_x and other pollutants, and so attention has more recently been redirected towards changes in engine design.

As with NO_x production in power stations and industrial boilers, rates of production of NO_x in conventional (petrol-based) vehicle engines depend largely on the amount of air present at the time of combustion. This can be altered by adjusting the air–fuel ratio, which is defined as 'the number of pounds of air used to burn each pound of fuel' (Lynn 1976: 248). The optimum efficiency in a conventional engine occurs with an air–fuel ratio of 15 (the so-called 'stoichiometric point'), but the ratio varies with mode of operation of the engine (maximum power for acceleration comes at a ratio of 12–13, economical cruising at steady speed has a ratio above 15).

Most early vehicle pollution control legislation in the United States (during the 1960s) focused on reducing emissions of carbon monoxide (CO) and hydrocarbons (the latter play a role in forming photo-oxidants like ozone, hence in forming acid rain – see Figure 2.6). This could be done by adjusting the air–fuel ratio towards the chemically optimum level of 15. But at that point increased combustion efficiency raises the temperature of combustion; consequently when CO and hydrocarbon emissions are low, the formation of NO_x is at a maximum (Lynn 1976: 248). As a result, nitrogen dioxide emissions increased while attempts were made (by optimizing engine efficiency) to meet the CO and hydrocarbon standards.

An alternative emission reduction approach had to be sought by the automobile industry when the first American NO_x standards were introduced, in California in 1971. It was based on recirculating part of the exhaust gases back into the engine and mixing them with the incoming air. As in the NO_x 'flue gas recirculation' option outlined above, this reduces the oxygen level in the combustion chamber, lowers the temperature of combustion and thus reduces the ultimate production of NO_x.

As vehicle NO_x emission standards in the United States became more stringent through the 1970s, further changes were required. These include more complex controls to vary the air–fuel ratio (especially towards more 'lean burn' mixtures, which have more air relative to fuel) and modified ignition systems. Moreover, additional devices had to be fitted to most cars. The most common was to replace the conventional exhaust muffler with a catalytic converter in the exhaust pipe, which would further oxidize the incompletely burned CO and hydrocarbons to CO_2 and water and thus reduce emissions.

Option 2. Change to a different type of engine

There are, of course, other ways of reducing NO_x emissions from vehicle exhausts (Lynn 1976). One strategy is to switch from the conventional petrol-driven (liquid gasoline) engine to other types of engine. One option here is to make more use of alternative forms of internal combustion engine, such as liquefied petroleum gas (LPG), which allows more efficient combustion and significantly reduces CO and hydrocarbon emissions but on its own would do little to solve the NO_x problem.

Another option is to use a rotary rather than the conventional internal combustion engine; the four-stroke Wankel engine is a possibility that would produce much less NO_x but more CO and hydrocarbons. Yet other options include the more widespread use of diesel and electric forms of vehicle power, and of novel forms of power, such as gas turbine or steam (Rankine-cycle) engines.

These types of alternative offer substitute methods of providing power in vehicles, but none looks likely to replace the conventional internal combustion engine as *the* main method, at least in the near future. A number of factors explain this apparent inertia, including economic and technical limitations of the main alternatives. There is also a marked reluctance amongst purchasers of vehicles (private and institutional) to think seriously about the alternatives, especially in a climate of free choice and what is assumed to be the unproven reliability and viability of most of the alternatives (the 'Why change if you don't have to' mentality).

Option 3. Transport planning

There is growing awareness that enlightened transport planning can make a substantial contribution to the reduction of NO_x emissions from vehicles. One means of achieving this is to improve the flow of traffic in congested areas; pollution levels are reduced if bottle-necks and traffic jams are minimized. NO_x emissions rise sharply as vehicle speed increases, so that traffic planning designed to cut down on high-speed vehicle movements (and proper legal enforcement of upper speed limits) would also make a contribution to the problem.

Transport planning is most successful in reducing pollution from vehicle exhausts when it seeks to cut down on the number of vehicles on the road in an area by providing 'mass transit' alternatives. The objective here is to encourage private car users (who often travel alone, or perhaps with one passenger – both extremely inefficient modes of travel) to use public transport systems, especially during rush-hour commuter trips to/from busy city centres. Obvious examples include the rapid rail transit systems in major US cities (such as New York, Boston, and San Francisco), the heavily used commuter trains in London, England, and limited stop commuter buses to/from many major cities in Europe and North America.

Providing the mass transit alternative to the private car is one thing; encouraging commuters to use it is another. Old habits die hard, and – like sluggish snails in their comfortable shells – commuters doggedly stick to their own mode of transport for comfort and convenience (despite its heavy relative cost). In some cities commuters are encouraged to switch to mass transit systems (or at least to car-share schemes) by various measures, such as heavy costs or limited availability of city-centre parking.

Priority schemes have also been talked about, in which access to a given part of a city is limited on certain days to certain cars. An interesting example of such a scheme in practice is Athens, Greece (Dicks 1985). The city has long suffered from thick smogs (*nefos*) which envelop it for much of the year. When *nefos* is particularly bad, private cars with odd and even number plates are banned from the city centre on alternate dates; private cars have also been excluded completely from central Athens for limited trial periods.

Such traffic management schemes have numerous advantages other than the reduction of NO_x emissions and prevention of severe pollution episodes. These include reduced overcrowding in city centres, reduced road repair and maintenance costs, reduced need to set aside land for car parking and for (multi-lane) highway construction, reduced accident rates and human stress, and decreased public travel costs. All in all, these are valuable savings, whether costed in purely economic terms or in terms of improved quality of life, and this option is likely to be pursued more widely in the future.

CONCLUSIONS

Prevention is better than cure, in pollution control as in other aspects of life. Whilst it is possible to 'cure' some of the worst biological effects of acidification by adding lime or limestone to affected lakes, soils, and forests, such remedial action offers at best a palliative. Liming does *not* offer a viable long-term solution. The only acceptable solution is to make sure that emissions of the sulphur and nitrogen oxides that provide the precursors to acid rain are reduced to acceptable levels.

This chapter has explored the main options for reducing emissions of SO_2 and NO_x from power stations, industrial point sources, and vehicles. We conclude that viable and technically effective methods of controlling and reducing such emissions already exist, and that further options – perhaps even more viable in economic book-keeping terms – are on the horizon. Such methods, applied separately or in combination (as circumstances dictate) can bring reductions to agreed levels within agreed time-scales. 'Energy conservation and the substitution of alternatives to fossil fuels must be combined with the desulphurization of fuels and stack gases in order to arrive at the optimum strategy', argued Goran Persson of the Swedish Environment Protection Board (Persson *et al.* 1976: 249).

Proof that a solution to the oxide emission problem exists, given political recognition of the problem and willingness to tackle it, comes from Japan (La Bastille 1981: 680). In 1968 the Japanese government issued strict SO_x controls and encouraged the use of low-sulphur fuels and desulphurization. By 1975 sulphur oxide emissions had halved, even as energy consumption doubled. Since 1975 even stricter controls have been set, and by 1982 nearly 1,200 scrubbers had been fitted in Japan (compared with around 200 in the United States).

The problem is thus *not* a technical one; it is essentially a political one (see Part IV). Unless there is genuine political goodwill (and the sustained public pressure to ensure this), the acid rain 'problem' – which *can* be solved – will *not* be solved. The central question is whether to spend the vast sums required to embark on long-term oxide emission programmes of the scale called for in the EEC Directive (i.e. a 60 per cent reduction in SO_2 levels by 1995).

But the high cost of cleaning up is in no sense 'money down the drain', because most if not all of the costs may be offset by positive side-effects such as the creation of new jobs and generation of useful by-products (such as commercial sulphuric acid) (La Bastille 1982: 680), not to mention the savings to be gained on corrosion of materials (chapter 6), the values of conserving fish (chapter 4), forests and crops (chapter 5), and the benefits in improved human health (chapter 6).

As we shall see in the next chapter, the reluctance of the United States and Britain to join the other '30 per cent club' countries and curb sulphur and nitrogen oxide emissions rests in part on the questionable non-linearity of response of acid rain to oxide emissions (see chapter 3) and in part on their failure to accept (or to be convinced by) available scientific evidence (see Part II) that acid rain is *actually* causing the damage to the environment widely attributed to it.

Part IV

THE POLITICS
OF ACID RAIN

8 INTERNATIONAL CONCERN AND INITIATIVES

The acid rain storm . . .

What is this, the sound and rumour? What is this
     that all men hear,
Like the wind in hollow valleys when the storm is
     drawing near,
Like the rolling on of ocean in the eventide of fear?
     'Tis the people marching on.
(William Morris, Chants for Socialists, 'The march of the workers', 1885)

Recent prolonged droughts in the Sahel and in Ethiopia in Africa, with untold and seemingly relentless human misery and deprivation and a massive human death toll, reinforce our heavy reliance on the rainfall that permits land to sustain life. Nowhere on earth can life survive for long without Shakespeare's 'gentle rain' which 'droppeth . . . from heaven' (*Merchant of Venice*, *IV*, i, 182). But it is not simply the *quantity* of rainfall that matters; the *quality* of that rain is highly important as well.

We have seen in Part II how rain that is more acid than normal can seriously affect rivers and lakes (and the life within them), soils, forests, crops, buildings, and structures, even the health and survival of humans. We have also seen (in Part I) how human activities cause at least part of the acidification of rainfall that causes these problems, and (in Part III) that there *are* techniques available that can reduce this acidification. Little wonder, therefore, that there is mounting concern to do something about the acid rain problem, and to do it without further delay.

In this Part we shall explore the roots of this interest and examine how it has emerged and been expressed, especially since the mid-1970s. We begin, in this chapter, with a general review of the international debate, before moving on to examine in closer detail the nature of the debate in the USA (chapter 9) and in Great Britain (chapter 10).

THE DEBATE OPENS – EARLY MILESTONES

As we noted early in chapter 1, acid deposition is *not* new. Wet deposition was first described by R.A. Smith in Manchester as far back as 1852, and dry deposition has long affected many of the world's major cities. What *is* new is the depth of interest in the issue, arising from the growth of wet deposition and the spread of acid rain over much of western Europe and North America (see Figure 1.1).

Scandinavian concern

Acid precipitation was long seen as a problem limited to Sweden and Norway (Paulsson 1984). Many lakes there were turning acidic and losing their fish populations in the 1920s and 1930s (see chapter 4), but the cause of the acidification remained at the time an unsolved mystery.

The possible link with acid precipitation started to emerge when first results appeared from the European Air Chemistry Network (EACN), which had been started in the 1950s by Sweden (by the International Meteorological Institute in Stockholm) (Oden 1976; Granat 1977). By 1955 it was clear (from EACN observations) that precipitation over large parts of Europe was unusually acidic (Egner and Eriksson 1955). Within fifteen years the data were to show that rainfall acidity was increasing (i.e. its pH was falling), and the area over which acid rain was falling was also increasing.

A major breakthrough came in 1968, when the Swedish scientist Svente Oden reported increased acidity of lakes in southern Scandinavia (which was seen to be closely associated with large-scale fish deaths (Oden 1976). But the main significance of Oden's study was that it showed that the most likely cause of the acidification was airborne sulphur that had been blown over to Scandinavia from other countries (mainly East and West Germany and Great Britain). This was the first tangible evidence that acid deposition is a serious international problem, but at the time the study was greeted with marked indifference by the scientific world and marked silence by politicians beyond Scandinavia. Only later was it to be hailed as a milestone in air pollution research, and given the wider recognition it so clearly deserved.

Watershed year – 1972

There was mounting evidence by 1972 (especially from EACN) that the precipitation in Europe was becoming increasingly more acidic, and it was becoming clear that *something* must be done. It was to be a year of significant developments, literally a watershed in the acid rain saga. The United Nations Conference on the Human Environment – itself a milestone in environmental diplomacy (Ward and Dubos 1972) – was held in Stockholm from 5 to 16 June. A wide range of environmental problems were discussed by delegates from many countries; this was the first serious international dialogue on many pollution issues. Sweden used the distinguished gathering as a platform from which to air its growing concern over acid rain and to launch its since oft-repeated call for proper control of long-range sulphur pollution. Few countries took the call seriously, especially those (like Britain and West Germany) who were accused as major sulphur emitters.

However, one tangible outcome of the Stockholm Conference was the unanimous adoption of twenty-six general principles relating to protection of the environment in and between UN member states. Perhaps the most important is Principle 21, which declares:

> States have, in accordance with the Charter of the United Nations and the principles of international law, the sovereign right to exploit their resources pursuant of their own environmental policies and the responsibility to ensure that activities within their jurisdiction or control do not cause damage to the environment of other states or of areas beyond the limits of national jurisdiction.
>
> (UN General Assembly, Document A/Res/2997, xxvii, 15)

The Organization for Economic Co-operation and Development (OECD) took an important initiative that same year by launching a co-operative programme to measure the Long-Range Transport of Air Pollutants (LRTAP) (Ottar 1976). It involved a network of about seventy ground stations, which took daily measurements of air quality (Ottar 1976); results were reported each month to the Norwegian Institute for Air Research, which took responsibility for analysing the results (Reed 1976). This was the first monitoring devoted specifically to acid precipitation, and it has yielded invaluable information.

However, simply measuring the acidity of rainfall does little to help understanding of how acid rain affects the environment. The Scandinavians had by 1972 collected sufficient evidence of acid damage to forests and freshwater lakes and rivers to realize that it was becoming an environmental problem of major significance, about which more should be known. Norway took the lead in starting a detailed research project on the effects of acid precipitation on forest and fish (Abrahamsen *et al.* 1976). The project (widely referred to as the SNSF Project) was funded jointly by the Norwegian Council for Scientific and Industrial Research, the Agricultural Research Council of Norway, and the Norwegian Ministry of Environment. It was initially planned to run from 1972 to 1975, but later extended to 1979.

Increasing awareness

Early results from the SNSF Project were presented at an International Conference on the Effects of Acid Precipitation, held in Telemark, Norway, from 14 to 19 June 1976 and attended by delegates from twenty countries and six intergovernmental organizations. The conference recommended that 'all governments reconsider their approaches to the control of the emission of relevant pollutants, bearing in mind the available range of technical solutions. A reduction of emissions would also have beneficial effects in areas close to the emission sources' (Anon 1976).

By 1977 there was no doubt that long-range transport of sulphur dioxide and acid pollutants did occur across Europe, and that this transport in upper air winds had no respect for national boundaries. OECD studies, under the LRTAP programme (Organization for Economic Co-operation and Development 1977), had established that the region receiving rainfall ten to thirty times more acidic than normal had grown to include the eastern half of Britain, most of central Europe, and Scandinavia up to north-central Sweden and Finland (see Figure 2.3). They had also established that some countries export acid rain, whilst others import it. For example, it was discovered that under certain conditions up to half of Britain's sulphur emissions (mainly from power stations) were exported to other European countries (who were, not surprisingly, not amused at their uninvited imports!).

Reliable evidence from North America has only been collected since 1977, when the Canadian Atmospheric Network for Sampling Air Pollutants (CANSAP) began; the US National Atmospheric Deposition Program started in 1978 (Whelpdale 1983: 73). Data from each reinforced the lack of respect of the oxide-bearing winds for political or administrative boundaries.

By the late 1970s there was also a better awareness that earlier air pollution control strategies in the UK and USA (the 'tall stacks policies' – see chapter 1) had been successful in reducing smoke problems, but – by dispersing gaseous pollutants downwind – had seriously aggravated conditions in remote areas up to thousands of kilometres away.

What was becoming apparent was that the problem of acid rain cannot be solved at a national level. As the Swedes had argued at Stockholm in 1972, whilst national initiatives can make significant contribution to reducing emissions and thus deposition rates, the problem is truly trans-national in character and scale. International co-operation is needed if the problem is to be resolved.

THE GAME PLAN UNFOLDS

Before we examine how this international co-operation began, it is useful to take stock of the implications of the evidence on precipitation acidity that had been collected from the monitoring programmes that began in the 1970s in Europe and North America. A note of caution is appropriate, because even by the late 1980s it is difficult to secure reliable direct measurements of rates of sulphur emission and deposition from many areas and some countries; the evidence is at best patchy and of variable quality (despite international schemes designed to harmonize procedures and instrumentation). The problem is understandable, given the massive number of small industrial and domestic sources of SO_2, the mobility of vehicle sources of NO_x and SO_2, and the variability through time in emissions from large sources of SO_2 (mainly power stations).

Differing estimates

A number of estimates of rates of emission and deposition have been made since the early 1970s, and often the figures are not identical from one study to another. This partly reflects the fact that rates of emission and deposition within a country change through time (see, for example, Figures 1.5 and 2.5). But it also reflects the fact that high-quality data, based on direct measurement, are simply *not* available for some countries; estimates are all we have to go on. The best information available to date comes from the United Nations Economic Commission for Europe (ECE) European Monitoring and Evaluation Programme (EMEP) (Dovland and Saltbones 1979), the World Meteorological Office Background Air Pollution Monitoring Network (BAPMON) (Georgii 1981), the Canadian Atmospheric Network for Sampling Air Pollutants (CANSAP), and the United States National Atmospheric Deposition Program (NADP).

Although the absolute values quoted for each country vary between surveys (e.g. the 1973 OECD results (see Table 8.1), Swedish estimates for 1980 – Swedish Ministry of Agriculture 1984, and the EMEP data for 1980 (see Table 8.1)), more times than not the countries maintain their positions relative to each other. In other words, the same countries tend to emerge as bad polluters in each study. Thus, for example, the UK, West Germany, and France remained the 'top three' group between the 1973 and 1980 estimates (Table 8.1), even though the absolute values of the estimates increased by more than two-thirds over the period.

The most comprehensive data are available for 1980 (although more recent data are available for *some* countries), and this is a convenient 'base

Table 8.1   *Changing estimates of sulphur emissions in Europe, 1973–80*

Country	Total emission of SO_2 per country (thousand tonnes) 1973 OECD est.	1980 EMEP est.	% change 1973–80
Austria	221	440	+ 99
Belgium	499	809	+ 62
Denmark	312	437	+ 40
Finland	274	595	+ 117
France	1,616	3,270	+ 102
Netherlands	391	480	+ 23
Norway	91	137	+ 51
Sweden	415	496	+ 20
Switzerland	76	119	+ 57
United Kingdom	2,803	4,680	+ 67
West Germany	1,964	3,200	+ 63
Total	8,662	14,633	+ 69

Source: based on Organization for Economic Co-operation and Development (1979) and EMEP data in McCormick (1985)

year' to use for comparisons. Table 8.2 lists estimated rates of SO₂ emission by country in that year (based on EMEP *measurements*), and Figure 8.1 shows emissions in Europe estimated by a modelling procedure. Again, absolute values varied, but relative positions remained relatively stable between the two. It is clear that there are two major emitters of SO₂ – Russia and the United States – which, between them, accounted for half

Table 8.2   *Estimated rates of emission of SO₂ by country, 1980*

Rank	Country	SO₂ emissions (thousand tonnes)	% of total deposition from within that country
1	USSR†	25,000	53
2	USA	24,100	?
3	China	12,000	?
4	United Kingdom*	4,680	79
5	Canada	4,516	50
6	East Germany†	4,000	65
7	Italy*	3,800	70
8	France*	3,270	52
9	West Germany*	3,200	48
10	Czechoslovakia†	3,100	37
11	Yugoslavia†	3,000	51
12	Poland†	2,755	42
13	Spain (1985)	2,730	63
14	Hungary†	1,633	42
15	Turkey	1,000	42
16	Bulgaria†	1,000	45
17	Belgium*	809	41
18	Greece*	700 (est.)	37
19	Finland	595	26
20	Sweden	496	18
21	Netherlands*	480	23
22	Austria	440	15
23	Denmark*	437	36
24	Romania†	200	36
25	Portugal	170	27
26	Ireland*	170	28
27	Norway	137	8
28	Switzerland	119	10
29	Albania†	100 (est.)	15
30	Luxembourg*	50	27
31	Iceland	10	0

Estimated total for EEC (*)	17.6 million tonnes
Estimated total for North America	28.6 million tonnes
Estimated total for Eastern Europe (†)	40.8 million tonnes
Estimated world total	around 100 million tonnes

Source: based on EMEP data in McCormick (1985)

Figure 8.1 Estimated sulphur dioxide emissions in Europe, 1980 (million tonnes per year)

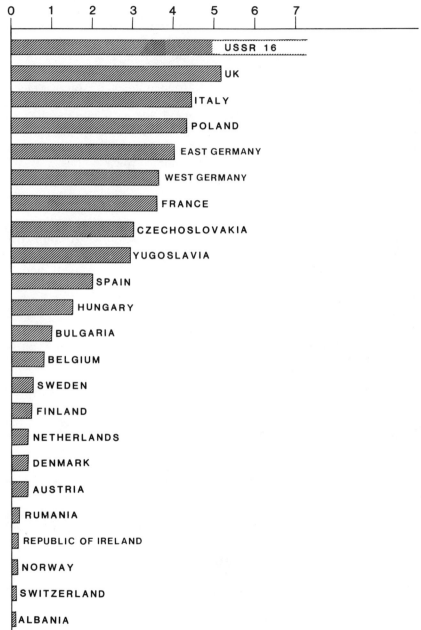

Source: based on data in Eliassen and Saltbones (1983)

of the world man-made output (Table 8.2). The 'top ten' countries accounted for nearly 88 per cent of the world total. Little wonder, therefore, that many countries look towards these heavy polluters to face up to their responsibilities and cut SO_2 emissions to reduce acid rain.

Give and take

The data show that most countries in Europe and North America receive acidified rain (defined as rain with a pH below 5.6 – see Figure 2.1), but that levels of acidification vary from place to place (Kallend *et al.* 1983) (see Figure 1.1) and through time (Rodhe and Grant 1984) (see Table 2.1 and Figure 2.2). In each country, at least part of the acid deposition comes from sources within that country – i.e. it is 'home-grown' pollution. But the sulphur and nitrogen oxides can travel long distances before they fall to the ground (see Figure 2.7), so that acid rain can be exported from one country to another (downwind).

OECD data show that some of the sulphur that falls on Europe's western seaboard might well have been transported across the Atlantic from eastern North America (*The Times*, 12 June 1984); up to a quarter of Norway's sulphur receipts are believed to come from Canada and the United States. Studies of Arctic haze in Alaska (chapter 3) also suggest long-distance transport of pollutants, perhaps from sources in northern Russia.

Consequently some countries tend to be net importers whilst others are net exporters of acid deposition. The clash of self-interest between the exporters (who are happy to get rid of the SO_2) and the importers (who are reluctant to receive it) is the central pivot of the acid rain debate.

The records show that Britain, West Germany, Italy, and several Eastern European countries export large amounts of sulphur pollutants, whilst the Scandinavian countries receive much more from abroad than they produce themselves (Figure 8.2). Many of the major importing countries are member states of the European Community (Table 8.3), and their growing frustration at taking in their neighbours' 'dirty washing' has acted as an important catalyst in promoting EEC initiatives in air pollution control.

But the tension between importers and exporters is not confined to Western Europe. A number of countries in Eastern Europe import most of their SO_2 from their neighbours (Table 8.3). In North America, large amounts of sulphur pollutants are believed to be blown into eastern Canada from the Eastern United States (see chapter 9), although it is clear that Canada does produce a fair proportion (around half) of its own acid rain (Table 8.2).

The major emitters (Table 8.2), and those who receive relatively little SO_2 from other countries (Table 8.4), tend to view the acid rain problem – especially from an international perspective – with much less urgency than those who import it. Recipients suffer the consequences (in terms of acid

Figure 8.2 Estimated rates and pattern of deposition of sulphur over Europe, 1980

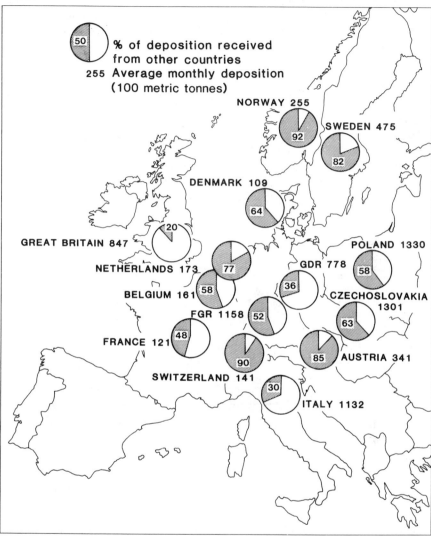

Source: based on EMEP/ECE data, after Environmental Resources Ltd (1984)
Note: Showing the estimated average monthly rate of deposition (in hundreds of tonnes) and estimated percentage of deposition that is imported from other countries. The percentage values shown differ slightly from those in Tables 8.2 and 8.3 because they are based on different data sets.

damage – see Part II) of what they see as unscrupulous exporting and immoral indifference by donors.

Table 8.3 *Major importers of SO₂, 1980*

Country	% of sulphur deposition imported from foreign sources
Switzerland*	78
Austria*	76
Luxembourg*	73
Netherlands*	71
Albania[†]	67
Norway*	63
Sweden*	58
Romania[†]	56
Czechoslovakia[†]	56
Finland	55
Denmark	54
Belgium	53
Poland[†]	52
Hungary[†]	52
Greece*	51
Canada	around 50

Source: based on EMEP data in McCormick (1985)
Note: the data in this table and Table 8.2 do *not* add up to 100% because some proportion of the deposition remains unaccounted for in each country. * EEC countries; [†] Eastern European countries (see Table 8.2).

Table 8.4 *Countries that receive least SO₂ from foreign sources, 1980*

Country	% of sulphur deposition imported from foreign sources
United Kingdom*	12
Spain	18
Italy*	22
Iceland	25
East Germany[†]	32
USSR[†]	32
Ireland*	32
Portugal	33
France*	34
Turkey	39
Yugoslavia[†]	41
West Germany*	45
Bulgaria[†]	47

Source: based on EMEP data in McCormick (1985)
Note: relevant data not available for USA. * EEC countries; [†] Eastern European countries (see Table 8.2).

INTERNATIONAL INITIATIVES

ECE LRTAP Convention (1979)

International initatives began in 1979, when the ECE – a different body altogether from the European Economic Community (EEC) – drafted a Convention on the Long-Range Transport of Air Pollutants. The LRTAP Convention was formally signed by thirty-five countries (Table 8.5) in Geneva, on 13 November 1979. It aimed to encourage a co-ordinated programme of pollution control in European countries, designed to reduce emissions of sulphur dioxide and thus to curb acid rain (for the benefit of all countries in Europe).

The Convention has an interesting background, because it grew out of an initiative by Soviet President Leonid Brezhnev at the 1975 Helsinki Conference on Security and Cooperation in Europe (Rosenkranz 1983: 5). Given this strong Soviet stake in its adoption, the apparent reluctance of Eastern European countries (including Russia) to ratify it, and their notable absence from the 1982 Stockholm Conference on Acidification of the Environment, seem to reflect a marked change of heart.

The Convention was designed as an 'outline convention', *not* a legally binding document detailing required action by the individual countries. It outlined the general responsibilities of the signatory states to put limits on trans-frontier air pollution (covering *all* types of air pollutants, not just sulphur), and reiterated Principle 21 from the 1972 Stockholm Conference on the obligation of states to ensure that activities within their own territories did *not* give rise to environmental damage in other countries. Signatory states agreed to five things (Swedish Ministry of Agriculture 1984):

(a) recognition that airborne pollutants are a major problem
(b) declaration that they would 'endeavour to limit and, as far as possible, gradually reduce and prevent air pollution, including long range trans-boundary air pollution'
(c) intention that they would use 'the best available technology which is economically feasible' to achieve (b)
(d) intention to act together to work out guidelines and strategies for better control of emissions and pollutants
(e) intention to develop technical and scientific co-operation (e.g. in monitoring effects and measuring emissions).

It was hailed at the time as a turning point in international air pollution control; Erik Lykke, Norway's Minister of Environment, added 'I'm praying that this Convention will bring a cleaner landscape 10 to 15 years from now' (La Bastille 1981: 685). But the agreement was soon to be heavily criticized as 'not sufficient' because it 'makes only vague recommendations' (La Bastille 1981: 685). It set *no* targets for emission

Table 8.5   *Commitment of countries to the 1979 LRTAP Convention, by start of 1987*

Signatory	Date of ratification	Date of accession	Promised SO_2 reduction from 1980
Austria	Dec 1982	June 1983	50% by 1995
Belgium	July 1982	June 1984	50% by 1995
Bulgaria	June 1981	June 1984	30% by 1993
Byelorussian SSR	June 1980	June 1984	30% by 1993
Canada	Dec 1981	June 1983	50% by 1994
Czechoslovakia	Dec 1983	Sept 1984	30% by 1993
Denmark	June 1982	June 1983	50% by 1995
Fed. Rep. Germany	July 1982	June 1983	60% by 1993
Finland	Apr 1981	June 1983	50% by 1995
France	Nov 1981	Mar 1984	50% by 1990
German Dem. Rep.	June 1982	June 1984	30% by 1993
Greece	Aug 1983		
Hungary	Sept 1980	Apr 1985	30% by 1993
Iceland	May 1983		
Ireland	July 1982		
Italy	July 1982	Sept 1984	30% by 1993
Liechtenstein	Nov 1983	June 1984	30% by 1993
Luxembourg	July 1982	June 1984	30% by 1993
Netherlands	July 1982	Mar 1984	40% by 1995
Norway	Feb 1981	June 1983	50% by 1994
Poland	Mar 1985		
Portugal	Sept 1980		
Romania			
San Marino			
Spain	June 1982		
Sweden	Feb 1981	June 1983	65% by 1995
Switzerland	May 1983	June 1983	30% by 1995
Turkey	Apr 1983		
Ukrainian SSR	June 1980	June 1984	30% by 1993
USSR	May 1980	June 1984	30% by 1993
United Kingdom	July 1982		
USA	Nov 1981		
Vatican			
Yugoslavia			
EEC	July 1982		

Source: Acid News (1985), volume 4, page 3

reductions, introduced *no* machinery for making compliance mandatory, and set *no* timetable.

The initial impetus was soon to be lost; countries started to realize the type and scale of sacrifices that would be expected of them if they pressed ahead with the initial intentions. There was to be a delay of three and a half years – three and a half years of growing frustration for Scandinavian countries and of heart-searching for many other signatories – before enough countries ratified the Convention, and it could come into force (in March 1983).

Stockholm Conference on Acidification (1982)

By early 1982 the ECE Convention still lacked the necessary twenty-four ratifications, so it had not yet come into force. Moreover, even if it *were* to come into force without further delay, there were still no concrete plans on the horizon to reduce pollution. Frustration was most acute amongst the Scandinavian countries.

The Swedish government took the lead, in June 1982, by hosting a large international congress on acid rain. It was held on the tenth anniversary of the 1972 UN Conference on the Human Environment. Further poignancy was added by the fact that it was held in Stockholm, scene of Sweden's 1972 call for international action to cut SO_2 emissions. The general aim of the Ministerial Conference on Acidification of the Environment was to heighten awareness of the whole acid rain problem. But it had two very specific aims – to encourage enough ratifications to bring the Convention into force, and to encourage national and international action to reduce SO_2 emissions and curb acid deposition.

The formal meeting was preceded by a gathering of international experts (scientists and technicians) who 'threw down the gauntlet' to the ministers and diplomats to think seriously about *action* with their three main conclusions (Swedish Ministry of Agriculture 1984). They agreed, first, that we *do* know enough about the nature and effects of acid rain to take remedial action – contrary to the outdated belief hung on to by many politicians that there were still too many unresolved scientific questions. Their second conclusion was that all measures that reduce total emissions *will* also reduce the damage now being done by acid rain to sensitive lakes and streams – even if the link is not linear. They also agreed that currently available control technologies *can* reduce emissions, and *at an acceptable cost* – the normal caveat added by political decision makers.

The Ministerial Conference was attended by representatives of twenty-two of the thirty-one member states of the UN Economic Commission for Europe who had originally signed the 1979 LRTAP Convention (Pearce 1982c; Rosenkranz 1983). Several representatives echoed Sweden's call for concerted international action to bring the ECE Convention into operation without further delay.

The meeting agreed that urgent action should be taken under the Convention on Long-Range Transboundary Air Pollution, including:

- establish and implement concerted programmes to reduce SO_2 and, as soon as possible, NO_x emissions
- use the best technology available that is economically feasible to reduce these emissions, taking account of the need to minimize the production of wastes and pollution in other ways

- support research and development of advanced control technologies
- further develop the North American monitoring programmes and EMEP, through better geographical coverage, improved data on emissions, standardizing sampling and measurement techniques, and improved modelling.

However, there were three significant features of the Conference that attracted particular interest in the media.

Perhaps the most interesting feature of the meeting was the largely unexpected about-turn in attitudes towards acid rain shown by West Germany. West Germany, one of the largest polluters in Europe (Table 8.2) and thus a notable *bête noire*, had long been on the receiving end of the Scandinavian campaign to limit oxide emissions. It had also presented a host of objections to, and been a somewhat reluctant signatory of, the ECE Convention in 1979. But now it emerged in support of the Scandinavian call for international action, and it took a visibly *active* stance and adopted a high profile. Interior Minister Gerhart Baum called on all the nations attending the meeting to fight air pollution *at source*, using the best available technology. He urged them to follow his country's example and equip new and rebuilt plants with best available pollution-control technology (flue gas desulphurization (FGD) – chapter 7). He also pledged his government's commitment to a 60 per cent reduction in SO_2 emissions from power stations and large factories by 1993. This 60 per cent target was to be achieved by a programme including introducing mandatory use of FGD systems on new and existing installations, amending the country's basic clean air regime, requiring the latest state of technology in all installations, and promoting an EEC-wide strict regulation on motor vehicle emissions (aimed at reducing NO_x by 50 per cent).

West Germany was the first major European polluting nation to join Scandinavian countries as an ally in their fight against trans-boundary air pollution and acidification of the environment. The primary reason for this sudden about-turn was the discovery of the serious damage being caused to their own forests (*Waldsterben* – chapter 5), so there was clearly more than a little self-interest given the economic significance of the country's forests and timber industry. The direct trigger for the change in heart was the growing influence and rising membership of the Green Party, reflecting a remarkable growth of strength of the environmental movement in West Germany. Acid rain 'became a potent political issue and in the elections which took place [early in 1983], hardened politicians swiftly announced their conversion in the face of persistent campaigning by the "green" party' (*Acid News*, 1984, 4, p. 3). This 'greening' of West Germany was a significant landmark in the development of public environmental awareness in Europe (Porritt 1984).

The second feature of the Conference was the marked absence of most Eastern European countries, despite the strong Soviet vested interest in adoption of the Convention (recall the Brezhnev 1975 initiative). East

Germany was the only major Eastern European polluter to attend. Bulgaria and three of the east's heaviest polluters – Poland, Czechoslovakia, and the Soviet Union (Table 8.2) – boycotted it altogether. Hungary and Romania – relative 'small fry' in the international league (Table 8.2) – were present but played no active role in debate. This exhibition of indifference alarmed and annoyed many delegates who wondered about the extent to which Eastern European co-operation in the ECE scheme could be counted on.

The third major feature – of equal international significance to the West German about-turn – was the self-imposed isolation of the United States and the United Kingdom. From the outset each country was clearly prepared to make few concessions; neither government sent its senior minister. The United Kingdom, represented by Giles Shaw (Parliamentary Under-Secretary of State for the Environment), argued that there was still too little certainty about the benefits of new control programmes to justify *new* control measures. The USA, represented by Mrs Kathleen Bennett (Assistant Environmental Protection Agency Administrator for Noise and Radiation), argued that further measures to reduce acid deposition should *not* be taken until ongoing research was completed. These two arguments – both essentially aiming to buy time – were to be rehearsed many times in public, at national and international meetings, by both governments in the years ahead (see chapters 9 and 10).

The final statement of the Conference was strong and positive. It concluded that further concrete action was required to reduce SO_2 and NO_x emissions, using the best available technology that was economically feasible. It also concluded that FGD had been proved as the best technology for SO_2 control and it should be used in new and rebuilt installations (especially power stations). Significantly, in light of the scientists' views, the final statement stressed that any reductions in emissions *would* benefit sensitive areas.

By now the sides were well and truly lining up for the public international debate. The good guys had distinguished themselves from the bad guys, and the battle was under way.

GROWING AWARENESS AND HEATED EXCHANGES (1983)

The public interest in acid rain, which had started in Stockholm in 1982, was given further momentum the next year with the discovery of damage (especially to trees) in many countries across Europe. By early 1983 the Swiss government was voicing alarm at evidence of damage to trees in the Swiss Alps, and immediately attributed it to acid rain (*Acid News*, 1983, 2, p. 14). Plans were swiftly drawn up for the Swiss Federal Institute of Forest Research to prepare a Swiss National Forest Inventory (1983–6), which would be updated every five to ten years. In West Germany, authorities were both amazed and alarmed at the speed with which forest death (*Waldsterben*) appeared to be spreading (see Figure 5.5).

Soon forest damage attributed to air pollution was being reported from

Austria (where around 11 per cent of the country's total forest area was showing signs of damage). Trees in France also appeared to be under attack, in an almost unbroken strip running from Brittany in the west to the Vosges in the east, also in central France and the Pyrenees. Symptoms were similar to those being reported from West Germany (see Figure 5.4), Switzerland, and Austria (*Acid News*, 1984, 2, p. 9).

Serious national attempts to curb emissions also began in 1983. Denmark's Socialistic Folkesparti proposed to the Danish parliament that acidification of the environment should be reduced (*Acid News*, 1983, 4, p. 8), mainly by installing FGD equipment in major power stations. It was estimated that this would add 2–3 per cent to electricity costs but reap much more in benefits from reduced corrosion, better harvests, and reduced human health problems. The proposal was rejected by the government, pending the outcome of studies then under way by the Environment Board.

Launching of LRTAP

By early 1983, twenty-four of the original thirty-five signatories had ratified the 1979 ECE LRTAP Convention (Table 8.5), which could then come into force on 15 March (*Acid News*, 1983, 1, p. 2; *The Times*, 15 March 1983, p. 13). Implementation of the Convention was to be overseen by an Executive Body (EB), which included representatives of thirty-four individual countries and the EEC that had signed it.

The first meeting of the EB was held in Brussels from 7 to 10 June 1983. Many participants were well aware of the inherent weakness of the Convention as originally formulated, in that it lacked specific pollution-control targets (by quantity and date). Some took the initiative and offered firm proposals (*Acid News*, 1983, 5, p. 12). The Swedish, Norwegian, and Finnish governments submitted a proposal suggesting that total emissions of sulphur from each individual country should be reduced by 30 per cent between 1983 and 1993. Canada, Denmark, Austria, Switzerland, and West Germany supported the three Nordic countries. Switzerland, Austria, and West Germany also submitted a proposal designed to reduce NO_x emissions, which was supported by Canada and the Nordic countries. No action was taken on either proposal. Some countries (e.g. the USA) expressed their general approval of the proposals but disapproved of specific numerical targets.

The EB emerged as a body willing to make general statements and to entertain acid rhetoric, but most reluctant to set specific targets for reducing trans-boundary air pollution. It was widely viewed as a toothless watchdog, and many environmentalists bemoaned its apparent inertia. Some were even angered at the low level of reductions called for in the Nordic proposal – Norway's Stop Acid Rain Campaign quickly announced that 'the Nordic governments are being too cautious. It is not sufficient to demand that European countries should reduce their air pollutants by 30 per cent in the space of ten years. If the fishkill in Scandinavia is to be stopped, acid rain *must* be reduced more drastically than this, and *at once* (*Acid News*, 1983, 4, p. 5). They quoted the 1981 OECD study, which showed that a 50 per cent reduction in Western Europe's SO_2 emissions was possible over ten years, at a cost of around US$12 per person per year (a 2.5–3.5 per cent rise in energy costs).

Publicity and awareness

Activities to build public awareness and vent public frustration also started in 1983, especially in West Germany where the ravages of *Waldsterben* were widely publicized. Early in the new year a new environmental action group (called 'Robin Wood') (*Acid News*, 1983, 2, p. 3) was formed with around 100 members (including former Greenpeace activists). Its aim was to carry out a range of activities designed to bring acid rain to the forefront of public interest, via the mass media. Their first major day of action was 21 February, with activities planned for five major German cities, including the occupation of power station flue stacks (in Berlin and Cologne) and the 'planting' of dead trees in town centres (in Kiel and Bremen).

Building public awareness was also a keen issue in Scandinavia, where many initiatives to bring acid rain issues before the general public were launched. In April, the Swedish Secretariat on Acid Rain produced a four-colour 'Noah's Ark' poster (arguing that 'if acidification is allowed to continue, we shall soon have to build a new Noah's Ark'). Its text was printed in English, German, Swedish, Dutch, and Norwegian, and it was to be distributed (free) widely (*Acid News*, 1985, 2, p. 2). Other free and widely circulated Swedish material included a general leaflet on acid rain aimed at foreign tourists visiting Sweden, which was produced by the National Swedish Environment Protection Board (in English and German), and an illustrated booklet 'Acidification – a boundless threat to our environment' produced by the Swedish Ministry of Agriculture (also in English and German) (*Acid News*, 1983, 4, p. 12). A Swedish film of acid rain ('Another Silent Spring') won prizes at large international film festivals in Czechoslovakia and the Netherlands, and it was subsequently released with soundtracks in English, German, French, Danish, Dutch, Italian, Greek, and Swedish (*Acid News*, 1983, 5, p. 9).

The International Youth Federation for Environmental Studies and Conservation (IYF) (fifty youth organizations concerned with nature conservation and the environment) declared 18–24 April to be 1983 International Acid Rain Week (*Acid News*, 1983, 1, p. 2; 1983, 2, p. 4). Their objective was to encourage groups to devote themselves to acid rain issues over that week and to carry out activities (e.g. article writing, public meetings) designed to increase general public awareness of the problems involved. The ultimate goal was to mould public opinion towards supporting control measures and to mobilize public support for letter writing, campaigning, and lobbying in order to provide the political momentum for acid rain to be taken seriously by national governments.

International debate

International debate continued in Brussels (19–20 April), when the European Parliament Committee on the Environment, Public Health and Consumer Protection held a public hearing on acid deposition (*Acid News*, 1983, 2, p. 2). It was addressed by a series of experts (in forestry, soils, and water) and by representatives from six organizations – the European Environmental Bureau (EEB), European electricity producers (UNIPEDE), Britain's Central Electricity Generating Board (CEGB), the Organization for Economic Co-operation and Development (OECD), the United Nations Economic Commission for Europe (UNECE), and the United Nations Environmental Programme (UNEP).

All agreed that the adverse effects of acid rain were clear, and the debate centred *not* on whether or not to reduce emissions but on how it should be done and who should pay for it. Views were strictly polarized. Industry and electricity producers (concerned only with the costs rather than the benefits of pollution control) failed to accept the need for fast action and pointed instead to the need for further research. The environmental organizations and most researchers took the opposite view: action was needed urgently (especially to reduce emissions by 50 per cent in the next five years), with energy conservation and use of alternative (low-sulphur) fuels as top priorities.

The debate and concern were not confined to Europe. Mustafa Tolba, executive director of UNEP, singled out acid rain as one of three key issues in his official annual UNEP report *The State of the World Environment 1983*, published to coincide with World Environment Day (5 June) (*Acid News*, 1983, 4, p. 2). He saw acid rain as a 'particularly modern, post-industrial form of ruination . . . it is as widespread and careless of its victims and of international boundaries as the winds that disperse it'. 'The only lasting solution', he argued, 'is to reduce the emissions of the pollutants in the first place.'

Later in the year, European farmers and forestry owners joined the growing chorus of groups requesting wholesale cuts in oxide emissions. A resolution to control pollutants at source was adopted unanimously by the 35th General Assembly of the European Confederation of Agriculture, held at Wiesbaden, West Germany, 26–30 September (*Acid News*, 1984, 1, pp. 14–15). In the same month, a consensus emerged at the European Community meeting in Karlesruhe, West Germany, that the EEC should take what action it could *without delay*, and that measures should include cutting emissions of SO_2 from power stations (*Acid News*, 1984, 1, p. 1).

ACTIONS AND DECISIONS (1984)

The momentum gained in the international acid rain debate during 1983 was to be taken much further in 1984. The first moves were taken in late January, when the European parliament was presented with a report on damage from acid rain prepared by its Environment Committee (*Acid News*, 1984, 2, p. 2). It stressed that most estimates of the cost of acid rain damage (e.g. the 1981 OECD estimates of £33–44 billion per year across Europe) erred on the low side because they did *not* take into account damage to forests (especially in West Germany). The report called for a programme of international action to combat the menace of acid rain, including strict emission standards, increased research effort, and a long-term commitment to reduce oxide emissions. It argued that the heavy cost of clean up should be met at least in part by imposing a special levy on those who produce the oxides (the 'polluter pays' principle).

Within days, in early February, the Parliamentary Committee of the Council of Europe, then in session in Strasbourg, was reading a report presented to it by its Committee on Regional Planning and Local Authorities. This recommended that the Council should prepare a proposal for a European Convention to determine limits for air pollution emissions (to be signed by non-member countries as well as by members), in order to strengthen the 1979 ECE Convention. It advocated the need drastically to reduce emissions of SO_2 (by at least 50 per cent) and NO_x (by 90 per cent). The report also urged further research and better information exchange between member states.

Early in the year, therefore, the stage had already been set for further dialogue on the trans-national implications of acid rain, and the calls for concerted efforts in Europe had begun. The urgency of the situation was to become even more apparent in the following months, as reports of acidification damage continued to come in from many countries – clearly the damage was not going to stop, even if momentum in coping with it was at times put into a state of suspended animation.

Damage spreads

By mid-1984, the state of West Germany's forests appeared to have deteriorated considerably. Unofficial estimates were suggesting that around half of the country's forest land was being affected, at a cost in the region of DM 7–10 billion per year (*Acid News*, 1984, 1, pp. 6–7). By now the detectable forest damage was also spreading through Scandinavia, where fish deaths and lake acidification had been the major headache for over a decade. Public concern was great in Sweden when it was revealed that up to 10 per cent of the trees along some parts of the country's west coast were showing similar signs of damage to those reported from central Europe (*Acid News*, 1984, 1, pp. 8–9). The possible economic consequences were not lost in a country in which 300,000 people are employed in a forestry industry worth over 50 billion Swedish kroner.

Similar signs were appearing in Norway, particularly in the south where up to 30 per cent of the trees in the county of Vest-Agder had damaged tips

(*Acid News*, 1985, 1, p. 3). Lake damage was also starting to emerge in Finland. About half of the lakes in the western part of Nyland province (west of Helsinki) either showed signs of acidification or had low buffering capacities and were under threat (*Acid News*, 1984, 4, p. 12). Tree damage was also reported in softwoods (mainly pines) along the west coast of Denmark (*Acid News*, 1984, 2, p. 10), and further south in Luxembourg (where the same report spoke of 19 per cent of the trees being damaged, 4 per cent badly). Further south again, some 34 per cent of the trees in Switzerland were found to be damaged (8 per cent badly), mainly in the southern Alps (*Acid News*, 1985, 1, p. 9).

Eastern Europe was not escaping the ravages of acid rain. Tree damage was described over an area of nearly 4,000 sq. km in Poland, and fears were expressed that this would rise to nearly 30,000 sq. km if Poland pressed ahead with its industrialization plans without pollution control (*Acid News*, 1984, 4, p. 5; Kasina 1980). Air pollution was particularly bad (ten times the world average) in Czechoslovakia, and by mid-1984 some 37 per cent of its forests were either dead or spoiled beyond repair (*Acid News*, 1985, 1, p. 9). The tree death toll in Czechoslovakia was expected to rise to 60 per cent by the year 2000 if no action was taken to reduce oxide emissions and thus curb air pollution.

Political steps

Unilateral decision making to reduce oxide emissions also started in 1984. This was a significant step forward for the individual countries with the vision and conviction to make their own commitments, and a strong catalyst for the international debate that was by now coming to the boil. In February, France announced its intention to reduce its emissions of SO_2 by 50 per cent by 1990 (over 1980 levels) (*Acid News*, 1984, 3, p. 6), echoing West Germany's earlier declaration of a similar 50 per cent reduction.

In Denmark, the official report of a Commission on Acidification specially appointed by Environment Secretary Christian Christensen proposed reducing Denmark's SO_2 emissions by 30 per cent (from 1980 levels) by 1995 (*Acid News*, 1984, 2, p. 10). This would cost an estimated 3–4 billion in Danish kroner 1986–95 (around 50 Danish kroner per year for the average family), and be achieved mainly by reducing the acceptable sulphur levels in heavy and light fuels and by fitting FGD technology to power stations. Norway's Environment Secretary Rakel Surlein announced her government's intent to reduce SO_2 emissions by 50 per cent, mainly by imposing tougher rules on use of high-sulphur fuel oils and by introducing catalytic exchange cleaning on all new motor vehicles (from 1988) (*Acid News*, 1984, 2, p. 11).

Switzerland, now alarmed over the extent of its own forest damage, aimed its control measures more towards NO_x from vehicle exhaust emissions – it imposed strict speed limits (effective from 1 January 1985) of 120 kph (kilometres per hour) on motorways (down from 130 kph), 80 kph on open highways (from 100 kph), and 50 kph in built-up areas (from 60 kph) (*Acid News*, 1984, 5, p. 11).

However, the willingness to act unilaterally was *not* shared by all European countries. For example, Belgium – which was seeing mounting evidence of acid damage and facing rising public interest in acid rain – had announced (on World Environment Day 1983) a programme to reduce SO_2 emissions by 30 per cent between 1980 and 1993; yet by the end of 1984 no concrete policy proposals had been forthcoming. Critics of this 'fence sitting' posture attributed the lack of action to the slow pace of government action, the strength of the country's pro-nuclear lobby, and the reluctance of industry and energy producers (*Acid News*, 1984, 5, p. 15).

By this time, the scale and significance of the acid rain problem in Europe were clear to all, even those countries reluctant to act directly in reducing oxide emissions. Also beyond doubt was the trans-frontier character of the problem, in which some countries gained (they exported unwanted air pollutants) whilst others lost (they imported them) (Figure 8.2). It was obvious to European scientists and politicians that concerted international action was now required. Two highly significant developments followed in March.

EEC Directive

On 1 March 1984 the Council of Ministers of the EEC Commission in Brussels adopted a framework Directive designed to reduce emissions of pollutants from large combustion plants and to regulate those from vehicle exhausts (Anon 1984/5f). The Directive required that there should be provision for Community-wide emission limits, based on the best available technology, not involving excessive costs, and taking into account the nature and qualities of the emissions concerned.

The first proposal (Commission of the European Communities 1984b, c) covered 'large combustion plants', which included power stations, metal manufacturing and processing, chemical works, the food industry, and waste disposal installations. It laid down specific limits for the emission of SO_2, NO_x, and dust from all new plants with a thermal output of 50 megawatts or more, to be applicable from January 1985. For existing plants it advocated the setting up of national objectives by each member state for the reduction of total annual emissions. The required reductions – 60 per cent for SO_2, 40 per cent for NO_x and 40 per cent for dust – were to be achieved by December 1995. These targets were similar to those already adopted in West Germany.

The vehicle exhaust proposal (Commission of the European Communities 1983, 1984a) argued that all member countries of the European Community should introduce regulations similar to those in the United States (the so-called US-83 regulations) for cleaning motor vehicle exhausts, by 1995 or earlier. It also argued that member countries should make unleaded petrol generally available from the begining of 1989 or earlier.

The proposals received a mixed reaction. Some countries heralded them as a realistic target. Even Britain saw them as a step forward; William Waldegrave told the House of Commons on 5 March that they 'provide a consistent framework for future proposals on control of emissions from specific industries which will take full account of the need to balance economic, technical and environmental factors' (European Communities Background Report ISEC/84/84). Others shared the view of the European Environmental Bureau, which welcomed the Commission's proposals but 'feared that [they] create expectations which it cannot live up to, even when fully implemented – the target and date proposed are too low and too late' (*Acid News*, 1984, 3, pp. 10–11). They favoured a 50 per cent reduction of SO_2 emissions within five years, then a further 50 per cent reduction in the following five years (i.e. a 75 per cent drop in ten years).

Ottawa Ministerial Conference

Also in March, Canada hosted the International Conference of Ministers on Acid Rain, which took place in Ottawa. Delegates attended from Austria, Denmark, West Germany, Finland, France, the Netherlands, Norway, Sweden, and Switzerland, as well as Canada (Anon 1984/5f; *The Times*, 23 March 1984). The air was thick with good intentions (for example, Norway presented a proposal to get international agreement on a 50 per cent reduction in SO_2 emissions by 1993 – *Acid News*, 1984, 2, p. 11), but caution was not in short supply. The ministers unanimously supported a communiqué urging *all* signatories of the Geneva LRTAP Convention to get down seriously to the business of tackling international air pollution by reducing trans-boundary fluxes of SO_2 and NO_x by at least 30 per cent by 1993 (from 1980 levels); the '30 per cent club' was formally born in Ottawa.

A prime purpose of the two-day Conference was to generate pressure of public and diplomatic opinion on both the United States and the United Kingdom to join the club. France and the Netherlands announced that they would join; the USA and the UK again refused (Table 8.5).

An International Acid Rain Week (2–8 April) was designated, as in 1983, by IYF. Its aim was again to create large-scale publicity on problems of acidification, with the hope of influencing decision makers to reduce emissions of the pollutant oxides. Many activities were organized in Europe, including demonstrations in Switzerland, Finland, and Britain. Thousands of postcards demanding rapid steps to reduce emissions were sent from Sweden, Finland, Norway, and the Netherlands to governments of the large emitter countries (*Acid News*, 1984, 1, p. 2; 1984, 2, p. 4).

Munich International Conference

Without doubt the highlight of 1984 was the multilateral ministerial meeting held in Munich from 24 to 27 June. Signatories of the 1979 LRTAP Convention and international government organizations were invited to attend; all accepted and ministers and officials from thirty-two countries were present.

The initative for the International Conference on Environmental Protection came from West German Chancellor Helmut Kohl (*Acid News*, 1984, 2, p. 8). This was a classic 'poacher turns gamekeeper' tactic, in which the country previously criticized as a major pollution exporter turned into a leader of the campaign for suitable international action. Herr Friedrich Zimmerman, West Germany's Minister of the Interior, commented during the opening session that 'next to the strengthening of peace, environmental protection is the most important task of our age' (*The Times*, 26 June 1984, p. 6). Growing awareness of the extent of forest damage in the country (especially from 1983 onwards) was an inevitable catalyst.

The meeting had two main aims – to exchange views and research findings on damage from acidification and experiences of how to combat air pollution at source. But there was also a third item on the 'hidden agenda', and that was to 'oil the wheels' of the international political machinery in the hope of speeding up progress on implementing the LRTAP Convention. In this latter aim the Munich meeting met with some success.

The political momentum was already under way, because ten countries (Canada, West Germany, France, Sweden, Norway, Denmark, Finland, the Netherlands, Austria, and Switzerland) had already committed themselves (at the March meeting in Ottawa) to a 30 per cent reduction in SO_2 emissions. Belgium, Luxembourg, and Liechtenstein announced at Munich that they would join the '30 per cent club' (*Acid News*, 1984, 3, p. 4).

Eastern European countries were split in their willingness to commit themselves to the 30 per cent target, which by then was being accepted as viable, if far from ideal, as a basis for reducing damage from acid rain. Some of the east's largest emitters of SO_2, including Czechoslovakia, Poland, Hungary, and Romania, were most reluctant to make such commitments, even though each admitted having evidence of acidification damage within its borders. Others were more forthcoming, largely reflecting their growing anxiety over the visible damage to lakes and forests (for example, extensive liming of fields was being undertaken in the USSR by this time – *Acid News*, 1984, 4, p. 5). However, their decision also highlights keen political astuteness in showing willing in the arena of international diplomatic debate at a time when the other 'super power' (the USA) was clearly dragging its heels and being plain uncooperative. The USSR, East Germany, Bulgaria, and the Soviet republics of Byelorussia and the Ukraine each committed themselves to a *qualified* acceptance of the 30 per cent target, by agreeing to cut their SO_2 *exports* (not emissions) by 30 per cent by 1993.

By the close of play at Munich, therefore, eighteen nations had undertaken to reduce their SO_2 emissions by 30 per cent by 1993, so were in the '30 per cent club' (Table 8.5). This was a great advance from the start of play (ten nations), but some of the hottest contestants had escaped unscathed. Five countries between them produce the bulk of Western Europe's SO_2 emissions – the United Kingdom, West Germany, Italy, France, and Spain (Table 8.2) – but the UK, Italy, and Spain refused to endorse the 30 per cent call and only West Germany and France were club members. However, over the preceding year, several countries had clearly begun to show much greater awareness of the problems of acid rain, and greater determination to act.

The UK argued that it would require a more reasonable time-scale, and it was not convinced that reducing SO_2 emissions was necessarily the best cure for the acid damage ailment (which might *not* even have been diagnosed correctly). The British statement to the Conference was unequivocal: 'we see no point in making heroic efforts, at great cost, to control one out of many factors unless there is a reasonable expectation that such control would lead to a real improvement in the environment' and 'Britain's neighbours cannot expect her to enter upon a crash programme while there is still so much that is speculative in the argument' (*The Times*, 26 June 1984, p. 6).

The United States was another escapee, arguing that not enough was yet known about the likely effects of a 30 per cent cut in emissions to embark on such a costly course of action. Both countries stressed that they had made substantial progress in reducing air pollution since 1970, and insisted that this be recognized and appreciated overseas.

Britain and the United States emerged from the Munich meeting as 'marked men', with most of the '30 per cent club' members insistent that these two major polluting culprits should face up to their international responsibilities and join the club. As we shall see in the next two chapters, they have managed to remain 'free' (i.e. outside the club) since then by a combination of stubbornness and political inactivity.

Growing frustration

Further pressure was brought to bear on those parties to the LRTAP Convention who had not yet made a commitment to reduce their SO_2 emissions when the Convention's Executive Body held its second meeting in September. Again it met with some success, because two more countries (Italy and Czechoslovakia) agreed to join the '30 per cent club' (*Acid News*, 1984, 5, p. 7). The UK and USA were notable in their persistent lack of willingness to join. Scandinavian frustration over Britain's entrenched attitude was clearly expressed when the Norwegian Society for

Conservation of Nature appealed directly to Britain to accept the 30 per cent reduction in SO_2 emissions, arguing that 'we cannot afford to wait while Britain delays measures for cleaning emissions' (*Acid News*, 1984, 4, p. 12).

Whilst these efforts to 'catch the big fish' were under way, attempts were also in hand to tighten up the pollution control agreements even further. The Swedish Ministry of Agriculture was handed a plan in October 1984 by the Swedish Environment Protection Board which aimed at a 65 per cent reduction of SO_2 and a 30 per cent reduction of NO_x between 1980 and 1995 (*Acid News*, 1984, 4, p. 10). The former target would be reached by energy conservation, sulphur limits on fuels, and tight control of emissions from coal-fired furnaces; the latter by greater use of unleaded fuel and fitting of catalytic converters to vehicles, and use of fluidized bed technology (see chapter 7), low NO_x burners, and flue gas denitrification in industry.

Promotion of public interest in and better understanding of acid rain, and mobilization of environmental activists in publicity campaigns, was also a hallmark of Europe during the closing stages of 1984. Over 400 delegates from environmental organizations in eleven West European countries gathered for a weekend conference of forest death at Bregenz in the Austrian Alps (*Acid News*, 1984, 5, p. 3). They planned various activities, including the 1985 International Acid Rain Week (15–21 April 1985) and demonstrations for World Environment Day (5 June 1985). The success of such campaigns to date was evident in the results of a 1984 survey in West Germany which revealed that 97 per cent of the population had heard of the country's 'dying forests' (*Waldsterben*), more in fact than those who could name the country's Federal Chancellor or President (Muller 1984: 2).

ENFORCEMENT AND PROGRESS (1985)

With so many important (and often successful) national and international initiatives taken during 1984, it was not surprising that 1985 was to be a somewhat quieter year on the acid rain front. It was to be very much a year of considered reflection by politicians in Europe and North America, during which negotiating positions were re-evaluated and emerging scientific evidence carefully examined.

Low profile politics

Diplomatic exchanges over acid rain continued throughout 1985, but the issue was handled in a much more 'low profile' manner than in 1984. In early March, Swedish Prime Minister Olof Palme deplored Britain's refusal to join the '30 per cent club', and urged all Nordic countries to increase political pressure on Britain to act against acid rain (*The Times*, 5 March 1985).

Conciliation and gentle persuasion dominated most international dialogue on acid rain. In mid-January, for example, West Germany's Chancellor Helmut Kohl, also commenting on Britain's continued refusal to join the '30 per cent club', suggested that 'we must discuss the situation with each other, as it is completely different in various countries. You [i.e. Britain] have hardly any dying forests or none at all. We have areas – most British people at least have heard of the Black Forest – where we face a complete catastrophe, and we must find a sensible middle line between ecology and economy. But the death of the woods is, of course, a special thing for us' (*The Times*, 18 January 1985, p. 7).

The ECE Executive Body, heavily criticized for its lack of success in its first two meetings and largely eclipsed as a power-house for exciting innovations by the 1984 Ottawa and Munich meetings, held its third meeting in Helsinki, Finland, in July 1985 (*The Times*, 10 July 1985, p. 8). Representatives of thirty countries attended. Environmental ministers of twenty-one of them (including most countries in Western and Eastern Europe, along with Canada) signed the so-called Protocol on sulphur emissions. The Protocol was a legally binding document by which these countries (the so-called '30 per cent club' – Table 8.5) promised to reduce their emissions or trans-boundary fluxes of SO_2 by at least 30 per cent as soon as possible, and by 1993 at the latest. In his opening address, Klaus Sahlgren (Executive Secretary of the UNECE) stressed that 'the Protocol is not an end in itself, but rather a means of turning the Convention into a reality' (Aniansson 1985: 29).

Fourteen countries were either unable or unwilling to sign the Protocol, but they did agree to sign a 'declaration of intent' expressing their will and intent to reduce their emissions, and their support in principle of the Convention and the endeavour of the Protocol signatories to achieve concrete results within a foreseeable future. Most of the fourteen (including Poland and Ireland) argued that economic difficulties – they simply could not afford the costly pollution control technologies at this time – prevented them from signing the Protocol straight away. Two – not surprisingly Great Britain and the United States – had a different defence. They argued that there were *still* too many scientific uncertainties, and more research was required. Britain also insisted that its SO_2 emissions had fallen by 42 per cent between 1970 and 1985, and by 25 per cent between 1980 and 1985, and that – even without signing the Protocol – an overall reduction of 30 per cent in SO_2 and NO_x emissions between 1980 and the late 1990s seemed highly likely (Aniansson 1985: 29).

NO_x and vehicle exhausts

By early 1985 scientists were starting to switch their interest from SO_2 towards NO_x, when some forest damage studies suggested that NO_x might play a more significant role than SO_2 in producing *some* forms of biological damage. This switch in scientific interest was quickly accompanied by a switch in the arena of the political debate, especially over suitable (and acceptable) control policies, because vehicle exhaust emissions produce most NO_x (whilst power stations and heavy industry – main targets of the earlier campaigning – produce most SO_2).

Vehicle exhaust controls (using technologies outlined in chapter 7) were introduced in a number of countries. Strong overtures were being made in Europe to adopt the tight US and Japanese emission limits. After all, the argument went, US NO_x limits had dropped by 75 per cent between 1971 and 1983 (*Acid News*, 1985, 3, pp. 10–11), emissions of pollutants from road vehicles in Japan fell by 90 per cent between 1978 and 1985, and today's Japanese vehicles are one-third more efficient than corresponding models in 1975 (*Acid News*, 1985, 2, p. 3). Despite such positive signs, the EEC Council of Ministers decided in March 1985 (*Acid News*, 1985, 2, p. 2) *not* to adopt such tight limits, but instead to allow countries to adopt varying methods to meet a staggered series of targets for vehicle emissions (applicable to large cars from 1988, medium-sized cars from 1991, and *all* new cars after 1993).

Such a cautious move found little support among Europe's environmental groups. Friends of the Earth (England) found it 'too little too late', and favoured adoption of the US or Japanese standards, based on the introduction of Toyoto-type technologies (using 'lean burn' engines with oxidative and catalytic converters at different air–fuel mixtures) (*Acid News*, 1985, 2, p. 2). The West German group BBU (Bundesverband Burgerinitiativen Umweltschutz) favoured voluntary speed limits of 100 kph on motorways and 80 kph on other main roads (*Acid News*, 1985, 2, p. 2). The European Environmental Bureau described the standards as 'too lenient', and argued that they would *not* produce 'the equivalent effect of the US standards on the European environment', as intended (*Acid News*, 1985, 4, pp. 6–7).

Different measures were adopted in different countries, to meet the EEC targets. The Austrian government, for example, agreed that lead-free petrol should be available at *all* filling stations in Austria from October 1985, and that US emissions standards should be applied within the country (*Acid News*, 1985, 2, p. 4). Financial incentives would be used to encourage adoption of the appropriate technology – a reward of 7,000 schillings to be given to the owner of any car that met the required standards on 1 October 1991. Liechtenstein favoured a more varied programme, including imposing speed limits (50 kph in built-up areas, 80 kph outside) from 1 January 1985, equipping cars with catalytic converters (to be exempted from tax), and improvement of public transport systems (*Acid News*, 1985, 2, p. 4). West Germany adopted a policy that insisted that, from 1989 onwards, all new petrol driven cars in the country would have to conform to the US-83 requirements (from 1988 for all cars with engines larger than 2 litres) (*Acid News*, 1985, 1, p. 6). It concluded that the best way of achieving this was to fit catalytic converters. Financial incentives to speed the voluntary adoption of this new technology included free road tax for the first four to ten years (depending on engine size) on cars that had it fitted, and tax reductions on existing cars with it already fitted.

In December 1984, the EEC environment ministers had agreed that unleaded petrol should be generally available in all member countries from 1 October 1989 (*Acid News*, 1985, 1, p. 6). Lead-free petrol was already available in West Germany and Switzerland in 1984; it became available in Denmark, Sweden, Belgium, and Luxembourg at the beginning of 1985 (*Acid News*, 1985, 2, p. 4).

Activism, awareness, and concern

Public actions designed to bring more public interest in and concern about acid rain made 1985 very much the 'year of the activist'. By then, many of the environmental organizations had a better sense of awareness of what would interest the media (TV, radio, and press) and of how to capture and sustain public interest. In the past, pollution had *not* been seen as a particularly newsworthy topic; but now the public appetite had been whetted. The interest was there – it simply had to be fed and nurtured.

February saw the first round of public actions in West Germany, when RWE – a large power company that generates 40 per cent of the country's electricity and pumps out around 0.7 million tonnes of SO_2 each year in doing so – was 'brought to trial' before a mock tribunal in Essen (*Acid News*, 1985, 2, p. 12). The company was incensed, the media were interested, the authorities were confused, and the public were amused! Further action was dotted through the year, to keep acid rain at the forefront of public awareness. In March, for example, four large smokestacks at the Shell oil refinery in Pernis, near Rotterdam, were occupied by ten activists from FoE Holland, who flew banners with 'Stop Acid Rain Now' and 'Profit is sour' slogans from the top (*Acid News*, 1985, 2, p. 12). They stayed aloft for several hours before descending; none was arrested.

The campaigning momentum that the environmental groups had gained at Bregenz the preceding September was maintained at a meeting of twenty-five representatives from West Europe's leading environmental organizations, held in London from 15 to 17 March 1985. A common campaigning strategy was formulated at the meeting, designed to focus efforts on the main issues and to strengthen alliances with other interest groups and with industries engaged in the production and sale of pollution-abatement technologies (*Acid News*, 1985, 2, p. 14). In many ways this marked the 'coming of age' of the acid rain public campaign; the penny had finally dropped that an effective environmental campaign could only be fought with strong co-ordination of efforts and with co-operation from other interested parties.

The main showpiece for demonstrations and activities was the third International Acid Rain Week (14–21 April), organized again by IYF.

Events of many kinds, involving the public as well as activists, were held in Finland, Sweden, Norway, Denmark, West Germany, Poland, Czechoslovakia, the Netherlands, Belgium, Ireland, England, Scotland, Wales, France, Austria, Italy, Portugal, Canada, and the United States (*Acid News*, 1985, 2, pp. 10–11). Most events got good coverage in the media and helped to bring the message home to the public. One particularly effective action was the releasing of balloons at emission sources in Wales, Scotland, West Germany, Austria, and the United States to symbolize the spread of acidifying oxides from identifiable point sources. In London and Holland fake parking tickets explaining the NO_x problem were issued to perplexed drivers in large numbers (30,000 in London alone).

The campaigning wheels were kept turning further at the International Citizens Conference on Acid Rain, held in the Netherlands from 20 to 25 May and organized by FoE International. Over seventy delegates from twenty-eight countries attended. It was truly international, because most West European nations were represented, along with Poland, Hungary, Canada, the United States, Hong Kong, the Philippines, Malaysia, Indonesia, Brazil, Argentina, and Mexico (*Acid News*, 1985, 3, pp. 15–16). The meeting agreed to the creation of an Air Pollution Action Network, designed to promote exchange of information and stimulate co-ordination of action between different countries. But its main tangible product was the decision to launch serious acid rain campaigns in the United States and the United Kingdom (the two largest contributors to acid rain amongst western industrialized nations).

European environmental activists were keen to take the battle on to British soil. After all, the UK – the largest emitter of SO_2 in Western Europe, which exports twenty-three times as much sulphur as it imports and refused to join the '30 per cent club' – was seen as the 'dirty old man of Europe' (*Acid News*, 1985, 3, p. 10). The UK campaign was to focus on the country's tourist industry (its second largest), and involved calling on all foreign visitors to boycott the UK as a holiday destination *until* it agreed to reduce dramatically its SO_2 emissions. Citizens throughout Europe were invited to send 50,000 pre-printed postcards (with the caption 'We love your country . . . but not your pollution') to UK hotels a.id tourist information offices, to register the point (*Acid News*, 1985, 4, p. 16).

CONCLUSIONS

Since the mid-1970s, acid rain has been projected to the forefront of international political debate in the west. Damage believed to be caused by acid rain has been discovered in more and more countries. Public interest in the issue has been sparked off by press coverage and by publicity events by activists. This interest reflects a growing public concern for the environment: a recent OECD opinion poll found that 62 per cent of Americans and 59 per cent of Europeans rated environmental protection higher than economic growth (*The Times*, 13 September 1985). These two forces have been important catalysts in encouraging more and more governments to realize the significance of the problem and to face up to their own responsibilities to act for the collective benefit.

The recent history of national and international interest in acid rain reveals some major areas of agreement and some equally major areas of disagreement. Few would disagree with the verdict that acid rain is one of the greatest environmental problems of our time, or with the observation that acidification (however caused) has created problems for lakes and rivers, forests, and structures in many countries. Neither is there any doubt now that upper air wind systems blow oxides and pollutants vast distances, across national boundaries and even wide oceans. There is a smaller measure of agreement over *what* has caused the acidification, and even less over the most appropriate measures for solving the problem.

The lack of unanimity in national views on acid rain has not been for want of trying. Since the mid-1970s scientific and political meetings on the subject have been held in many countries; the scientific community in particular has devoted much research to the key ingredients in the 'acid rain problem' – from emission, through dispersion, to deposition; examining biological changes in freshwater, forest, and agricultural ecosystems (see Part II); devising and testing various control technologies for reducing SO_2 and NO_x emissions at source (see Part III). The scientific literature on different acid rain themes is now immense.

There is no escaping the fact that some scientific questions are as yet far from fully answered. For example, by late 1985 it had become apparent that sulphur *might* be less important as a cause of some forest damage than nitrogen oxides (the true significance of hydrocarbons was also under question by then). If that conclusion turns out – with the passage of time – to be true, then efforts to reduce biological damage should be directed more towards reducing vehicle exhaust emissions of NO_x than power station and industrial emissions of SO_2 (hence the interest in exhaust controls during 1985).

The key area of uncertainty remains linearity of the association between oxide emissions and acid deposition (wet and dry). If the association was linear, a given percentage reduction in oxide emissions would produce a proportional decrease in acid deposition, and politicians could take comfort from the fact that heavy investment in pollution control would be likely to produce tangible improvements in air quality (i.e. reductions in air pollution). But scientists have yet to establish that it *is* linear. Whilst this uncertainty remains, it is all too easy for politicians to refuse to accept rigid targets for emission reductions (like the 30 per cent drop in SO_2 emissions called for by ratifiers of the ECE LRTAP Convention).

There is uncertainty over another 'linearity', too; that is the linearity of the association between acid deposition and biological damage. Put simply, if that association is *not* linear (and there is no unequivocal evidence that it *is*), then a given reduction in rates of deposition may bring little real reduction of damage in acidified lakes or forests. Uncertainty over these two 'linearities' further complicates an already complicated debate.

We have seen how many countries now accept that SO_2 emissions from power stations play a significant part in producing acid rain, and how strong recommendations for firm emissions reductions within specified time-periods have come from Scandinavia and West Germany. The sudden about-turn in West German views – previously strongly against costly pollution control but now leading the campaign in favour of it – once serious forest damage had been established within that country, offers a salutory reminder of the power of self-interest in international politics (environmental or otherwise).

However, not *all* countries agree. Many of the Eastern European countries, some of them very heavy producers of SO_2, have failed to accept the need for emission controls – some (like Poland) because they simply cannot afford it at present, but others because they do not see it as a high priority area of investment or east–west co-operation. The USSR has added an interesting twist to the normal super-power 'cat and mouse game' by joining the so-called '30 per cent club', leaving the United States (who still refuses to join) looking intransigent and uncooperative in this arena of international politics. The United Kingdom and the United States – two of the largest producers of air pollution – have largely resisted international scientific, diplomatic, and political encouragement to cut their emissions of SO_2 and NO_x. The main platform of their defence is that present scientific understanding of acid rain is still too limited, and much more research is required before they could embark on costly pollution control programmes.

Acid rain provides a fascinating illustration of international politics at work. All the usual ingredients are there – hidden agendas, vested interests, frustration and elation, inspired initiatives, and raging rhetoric. Money and morality, as ever, are key forces in the political debate.

Money is inevitably important, because pollution control on the scale favoured by some countries requires vast investment in control technologies and the money has to be found from somewhere. Many politicians have eagerly argued that we cannot afford such costly clean-up programmes, either because the money simply isn't there or – more realistically – because it is ear-marked for more high priority uses (like defence and investment in nuclear power stations).

However, the *real* question is *not* whether or not society can afford to develop emission control systems, but whether or not we can afford *not* to control emissions! There are two issues intertwined. First, the biological defence. It is questionable whether many freshwater and forest ecosystems – on which we rely for resources and thus survival – can continue to function if acid deposition continues at today's levels (i.e. if emissions are *not* controlled). Second, the economic defence. Although acid rain is a multi-million dollar problem, savings against reduced repair costs and so on would be offset against costs. The problems of drawing realistic comparisons between costs (easier to quantify) and benefits (more difficult) are widely recognized, but a 1981 OECD survey concluded that benefits would outweigh costs.

Morality creeps into the debate at all levels. Some countries clearly have more than self-interest at heart, and they are prepared to make some sacrifices for the collective benefit – i.e. they will accept the cost of pollution control within their country because it will help to keep the environment at large (theirs and others') clean and biologically stable. Others place altruism low on their list of priorities.

Overlying this moral foundation is international law, recognized by all civilized states. The legal maxim *sic utere tuo ut alienum non laedeas* ('use your property so as not to injure your neighbours') is a general principle of international law (Johnson 1976). Whilst there is no general prohibition to pollute, states *cannot* simply do as they please. Their actions *should* observe three duties (Courage 1983): *due diligence* ('they are obliged to observe all care so as to prevent the pollution of areas beyond their jurisdiction, and to combat it when it occurs'), *good neighbourliness* ('according to which every state is forbidden to allow its territory to be used in a way prejudicial to the rights of other states'), and *solidarity* (under which 'states are obliged to take the interests of other states into account when engaging in or allowing emissions that prevent the use of air for other purposes').

The 'bottom line' is self-interest. Quite simply, the countries that would benefit most from reduced pollution are *not* the ones that would have to bear the highest costs. The two countries that have swum most aggressively against the tide of international opinion are the United States and the United Kingdom. It is fitting that they are each examined in some detail in the last two chapters.

9 ACID RAIN IN THE UNITED STATES

Gripes of wrath . . .

Our scientific community is still unclear as to . . . what control methods should be used. In our opinion, we just don't know enough yet to impose control measures at great cost to the American people with questionable results.

(A. Alan Hill, quoted in La Bastille, 1981: 676)

It is clear from what we have seen in the last chapter that the United States shares centre stage with the United Kingdom in the arena of international political debate over acid rain. Four factors won the US this prominent – if totally unwanted – position in the debate, and taken together they highlight the seriousness of the problem on that side of the Atlantic. One is the simple and largely uncontested fact that the United States emits vast amounts of SO_2 (around one-quarter of the world total – see Table 8.2), so it has a pre-eminent position in the league of 'bad guys'.

A second factor is the large amount of damage from acidification (however caused) now evident in North America. Signs of serious damage to lakes and forests started to emerge in the Eastern United States and in eastern Canada in the mid-1970s. Since then, damage has been reported from an ever-growing area, even extending over to the Midwest and the Southwest United States.

Factor three is the two-dimensional nature of the North American acid rain debate, with major conflicts on the east/west and the north/south axes. The first reflects the growing frustration of states in the Northeast United States that receive large quantities of acid deposition, which they argue comes from large industrial areas further to the west. The other dimension stems from the strong Canadian disenchantment at importing acid rain from across the border in the Northeast United States.

The fourth, and doubtless the most important, factor is the continued reluctance of the US government to commit itself to an internationally acceptable reduction in oxide emissions within an agreed time-scale. Alan Hill, quoted above, was speaking as Chairman of the President's Committee on Environmental Quality, and he voiced the opinion that many US politicians have held through the entire debate. The US remains (in late 1987) stubbornly outside the '30 per cent club' (Table 8.5), even after the UK – its 'partner in crime' – reluctantly conceded in September 1986 to a 14 per cent cut in SO_2 emissions by 1996 (see chapter 10).

ROOTS AND ROUTES

Air pollution has long posed serious problems in North American cities. The American people have grown accustomed to New York's grimy and polluted air, thick photo-chemical smogs in Los Angeles, and heavy palls of smoke and grey skies over major industrial centres. Over the post-war era the US government has launched many major initiatives designed to curb air pollution and clean the nation's skies – the annals are littered with legislation to control pollution from industry, power stations, vehicles, and other sources. Many of these have been remarkably successful, and many major cities now boast more sunshine, lower health bills, cleaner buildings, and generally improved conditions since pollution control began.

A major catalyst for recent change was the National Environmental Policy Act (NEPA), passed by Congress in 1969 under the Nixon administration (Marcus 1986: 52). The Act declared a national policy designed (amongst other things) to 'encourage . . . harmony between man and his environment; [and] to promote efforts which will prevent or eliminate damage to the environment . . .'. The Environmental Protection Agency (EPA) was created to bring the Act to life. Environmental legislation bloomed during the 1970s, fuelled by public pressure to combat declining environmental quality, inspired by NEPA, and steered by the EPA. Federal laws were passed on a wide range of issues – from human use of nature, through pollution control, to protection of endangered plant and animal species (Marcus 1986: 61). Especially relevant here was the replacement of the 1967 Air Quality Act in 1970 with a Clean Air Act (amended in 1973, 1974, and 1977), which was designed to protect and enhance the nation's air quality, set standards for designated pollutants, and establish visibility standards.

Much of the early thinking was based on the notion of dispersing pollutants through the environment, in order to decrease concentrations below levels that were found to be unacceptable or produce serious local problems. This was the basis of the 'tall stacks policies' (described in chapter 1), introduced in the 1970 Clean Air Act. But scientists have since realized that 'what goes up must come down', even if it does so further afield (downwind). With this has come a realization that long-distance transport of SO_2 and NO_x in upper air wind systems increases the prospect of turning dry gases into acids (see Figure 2.7). To a large extent, therefore, the legislation effectively converted a local problem of dry deposition into a more regional – and ultimately trans-frontier – problem of wet deposition (acid rain).

Dawn of awareness

Up to the early 1970s, acid rain was seen largely as a problem confined to Scandinavia. At the 1972 UN Conference in Stockholm (chapter 8), American delegates listened to Swedish reports of acidification damage with detached academic interest, and took little interest in the then quiet calls for proper control of long-range sulphur pollution.

The new era dawned on 14 June 1974, when a report appeared in the journal *Science* suggesting that acid rain was falling on the United States (Likens and Bormann 1974). Drs Gene Likens and Herbert Bormann – respected biologists from Cornell and Yale Universities respectively (see chapter 1) – argued that rain falling on parts of the Eastern US had increased in acidity to 100–1,000 times normal levels as a result of increased SO_2 and NO_x emissions, mainly from electricity and industrial plants. Previously, they argued, solid particles (i.e. soot) would have neutralized these toxic emissions; this buffering capacity (see chapter 3) disappeared with the installation of devices to eliminate these solid particles in smokestacks.

The *Science* report rang alarm bells and looked like a good story for the national media. The *New York Times* (*NYT*), for example, reported (13 June 1975, p. 1) that improvements in emission control in cities had been offset by rapid industrialization of rural areas (hence SO_2 emissions had increased by over 45 per cent). It also outlined the belief that increased acidity of rain was linked with reduced forest growth in New England and Sweden, with increasing lake acidity in Canada, Sweden, and the United States, and with acceleration of corrosion damage to man-made structures. This was the first shot in what was to become a prolonged and at times acrimonious battle to raise US awareness of and concern about acid rain.

It was to be some time before the nation was to embrace the acid rain issue with any sense of urgency or need (*NYT*, 4 July 1975, p. 10). But scientific interest was by now awakened, and within two years the three main scientific ingredients of the debate had been established in the US. One was the very *presence* of acid rain; the *Science* report was the first of a number that described high acidity in rain falling over the eastern side of the nation. The second was firm evidence of long-distance transport of the culprit oxides, and this was provided in plume-tracking experiments. An early example was a manned balloon flight from St Louis (Indiana) in June 1976, which showed that polluted air (in this case derived from a Shell Oil refinery at Alton, Illinois) can drift over 150 miles with little dilution (*NYT*, 23 June 1976, p. 27). The third ingredient was reliable evidence of acid damage. Early in 1977, surveys by a team from Cornell University, led by Dr Carl Schofield (*NYT*, 28 March 1977, p. 31; Schofield 1976), showed that water in over half of the high-level (over 2,000 ft) lakes in the Adirondack Mountains in northern New York State was then acidic (see Figure 4.2c), and 90 per cent of the lakes were devoid of fish (see Figure 4.5c). Since then the Adirondack lakes have yielded much evidence of the possible consequences of freshwater acidification, and they have figured large in the public and political debate in North America.

By the mid-1970s it was clear that the quality of the American environment was in decline, even with the legislation that had emerged since 1969. A National Wildlife Federation (NWF) report, published in January 1976 (*NYT*, 19 January 1976, p. 13; 16 January 1976, p. 28; 14 February 1976, p. 24), noted that, although air pollution overall appeared to be on the decline, many states had failed to meet the requirements of the federal air quality standards (laid down in the 1970 Clean Air Act) by the mid-1975 deadline. It was also becoming clear to a growing number of people that Congress should, as a matter of some urgency, amend the Act to prevent the serious deterioration of clean air where it then survived – in national parks and wilderness areas, and around national monuments.

One tangible outcome of this better awareness of the need for strong environmental planning was the emergence of public environmental awareness. This was reflected in a massive growth in membership of environmental groups (like Friends of the Earth). An opinion survey of conservationists, carried out by the National Wildlife Federation early in 1977, found that they listed air and water pollution as the most urgent environmental problems of the day (*NYT*, 26 March 1977, p. 34). It was also reflected in the emergence of environmental issues as key components in the political electioneering machinery. Pollution was identified as a major area of public concern that should be addressed by presidential aspirants in the 1976 elections (*NYT*, 28 March 1976, p. 14). On 4 July 1975 – an apt date – President Ford had outlined the philosophy of the government on which he would be seeking full-time re-election to the White House in 1976 (*NYT*, 4 July 1975, p. 10). He pledged total commitment to continued progress in cleaning up air and water, but also stressed the need to create new employment prospects and pursue continued economic progress.

Despite the glowing rhetoric, the Ford administration was to be a term of hesitation and reluctance in bringing environmental reforms and legislation. It was an era of backsliding on environmental regulation. For example, less than three weeks after his 4 July speech, he announced the intention to further delay (until 1982) the introduction of statutory emission standards on vehicles, originally intended for 1978 models (*NYT*, 25 July 1975, p. 9). New proposals approved by the Senate Committee on Public Works in February 1976 (*NYT*, 7 February 1976, p. 24) required that 10 per cent of 1979 cars meet the highest emission control tests, and allowed the other 90 per cent to continue to meet the 1978 standards up to 1980. In a message to Congress in January 1977 (*NYT*, 8 January 1977, p. 25), President Ford suggested amending the Clean Air Act to lower vehicle emission standards, allow more use of coal in power plants, and remove requirements preventing significant deterioration of air quality in areas that already met the standards. A lowering of expectations was also apparent when the EPA abandoned its campaign to impose mandatory transportation-control plans on New York City and other big US cities, as required by the Clean Air Act (*NYT*, 14 January 1977, p. 3). It favoured instead a reliance on *voluntary* compliance, because of the belief that local opposition to *mandatory* schemes – by now strong – would only produce endless (and costly) litigation.

Calls for action

By the mid-1970s calls were being made to alter the earlier 'tall stacks' policies and to make more use of the emission control technologies then available. An editorial in the *New York Times* on 14 January 1976 (p. 12) called for greater use of scrubbers (see chapter 7) to achieve clean air standards, and noted that over the preceding two years the total number of scrubbers installed in the US had increased from 44 to 118. This, it argued, had made it possible for many users to use high-sulphur fuels *without* corresponding increases in oxide emissions (beyond the clean air standards).

A feature of this time was the growing chorus of calls to amend the 1970 Clean Air Act, which by now was being widely violated and whose implementation in all states and cities was proving rather difficult. Congress considered a list of proposed amendments to the Act in 1976, but cost factors prevented them from adopting any (*NYT*, 9 January 1977, p. 39). Industry was calling for relaxation of some of the more costly or inhibiting clauses in the original Act. Early in 1977 the National Association of Manufacturers recommended changes that would allow industrial pollution up to legal limits in areas where air was still cleaner than existing limits, and further changes to permit additional pollution in areas where national pollution limits had already been exceeded (*NYT*, 27 January 1977, p. 45). The Association also called for removal of the requirement that scrubbers be used to combat industrial air pollution, to allow polluters to select other methods (which might be more cost-effective for them and still produce the same net improvement in air quality). When Jimmy Carter was elected to the White House in 1977, he pledged support for changing the 1970 Clean Air Act.

By this time, too, tensions were starting to appear between US states over the inter-state transfer of air pollution (and attendant damage). It had been known for many decades that the dominant movement of air masses over the United States was from west to east (Figure 9.1), but the implications of this for dispersal of air pollution were becoming painfully clear. There was, for example, mounting concern in New York City and New Jersey state – which by then had some of the worst polluted air in the United States – that they were becoming victims of pollution originating elsewhere. A 1975 report from the Boyce Thompson Institute (*NYT*, 13 October 1975, p. 33) blamed most of the air pollution in Northeastern states on smokestacks in Birmingham (Alabama), Pittsburgh (Pennsylvania), and Gary (Indiana). Fears were also being voiced that high acidity being found in rainfall samples from New Jersey *might* endanger crops within the state (*NYT*, 31 August 1976, p. 59) – with economic as well as biological consequences.

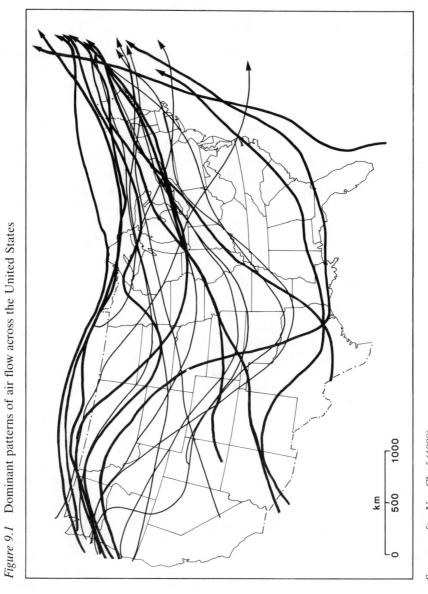

Figure 9.1 Dominant patterns of air flow across the United States

Source: after Van Cleef (1908)
Note: Based on an analysis of 1160 low pressure centres over the period 1896–1905

However, this was a period of serious national economic recession, and the cost implications of environmental protection were being keenly debated. EPA Administrator Russell Train estimated in mid-1976 that it would cost oil-refining and electricity companies over $30 billion over the next ten years to comply with *existing* pollution rules (*NYT*, 6 June 1976, p. 51). The US Commerce Department estimated that the US business community would have to spend $7.5 billion on new plant and equipment in 1977 alone to combat pollution (*NYT*, 25 May 1977, p. 5). There were some doubts over whether the ailing US economy could afford to shoulder such costs, and some calls to relax environmental controls – if only for a specified period – to help the economy.

There was also a strong and growing feeling that unless equally tight air quality standards were introduced in *all* states, those with the lowest standards would be at a competitive advantage in attracting new industries and keeping established ones (which would not be obliged to meet the high costs of installing and operating effective emission control technologies). Again the problem centred on New Jersey, which had much higher standards than those set under the 1970 Clean Air Act (and 1973 amendments) by the EPA and widely adopted in other states (*NYT*, 11 January 1976, p. 33; 16 January 1976, p. 63; 17 January 1976, p. 61). Early in 1976 studies were carried out in twelve areas within the state to determine whether state standards could be lowered in an effort to keep food-processing and glass industries from leaving the state. One option considered was to allow the burning of high-sulphur coal for generating electricity in the south of the state. On 12 February 1976 New Jersey Governor Byrne stressed the need for economic recovery – where necessary above the need for environmental improvement – in announcing the state's intention to relax air pollution emission standards for certain areas of south Jersey to permit industries and power companies to burn fuels with a higher sulphur content, until summer 1977 (*NYT*, 12 February 1976, p. 65).

CARTER INITIATIVES

The Carter administration (1977–81) inherited many of the serious economic problems that Gerald Ford had wrestled with, and it had a similarly mixed success in dealing with environmental problems.

Rising energy costs during the 'energy crisis' forced Jimmy Carter to formulate an Energy Plan during his first days of office in 1977. The plan involved – amongst other things – making more use of the nation's abundant coal supplies (*NYT*, 1 April 1977, p. 1; 23 March 1977, p. 4; 31 May 1977, p. 15). But the administration and its supporters, as well as its critics, were well aware that one result of reducing energy costs by using more domestic coal and less imported oil would be increasing emissions of SO_2 and NO_x, thus increased air pollution. The plan recognized the need to depart from pristine clean air standards (the sacrifice to be offered for the prospect of more and cheaper energy), and it included provision for financial incentives to help power plants to install pollution-control equipment when they switched from oil or natural gas to coal as a main fuel source. The requirement that new coal-burning boilers be fitted with flue-gas scrubbers (FGD technology – see chapter 7) was one of its most costly and controversial proposals.

However, the plan caused concern amongst US scientists. The National Academy of Sciences had earlier warned that burning fossil fuels was having serious adverse effects on the world's atmosphere, and the planneu increase in use of coal in the US was greeted with some anxiety (*NYT*, 31 July 1977, p. iv). Scientists at the Brookhaven National Laboratories advised that the plan could shorten many people's lives by up to fifteen years because of air pollution leading to chronic respiratory disease (*NYT*, 17 July 1977, p. 10). They forecast a toll of up to 35,000 premature deaths a year east of the Mississippi by the year 2010.

The Carter administration also inherited its predecessor's commitment to economic recovery, where necessary at the expense of environmental quality. Debate was heightened over whether the federal government should enforce strict pollution standards at the possible expense of industrial growth. The administration – faced with pro-relaxation lobbies on several fronts – supported comprehensive amendments of the 1970 Clean Air Act, particularly the movement away from compliance with tight emission standards. The coal industry (via the National Coal Association) was pressing for lowered air pollution standards to encourage electricity companies to burn more coal (*NYT*, 20 April 1977, p. 18) – partly to support the Energy Plan but mainly to boost their profits.

Vehicle emission standards were also high on the list of priorities for change. Vehicle manufacturers – particularly Ford, Chrysler, and General Motors – were keen to see standards relaxed, because otherwise cars then being made would be in violation of the 1970 schedule for reducing harmful emissions (*NYT*, 10 February 1977, p. 19; 23 April 1977, p. 18; 30 July 1977, p. 8; 3 August 1977, p. 11). The car industry had a powerful bargaining position in arguing that unless controversial amendments to the 1970 Act were agreed upon, they would not be able to begin legal production of the 1978 model cars so they would quite literally have to shut down the industry and lay off vast numbers of workers. Their strong hand won the day, because it was agreed to include in the Bill a two-year extension of existing tailpipe exhaust standards for three key pollutants (*NYT*, 4 August 1977, p. 1).

The House of Representatives, in a lengthy debate, considered many proposed changes to the 1970 Act and on 27 May 1977 it passed a Bill (326–49) formally amending it (*NYT*, 27 May 1977, p. 1). That Bill was passed by Senate on 5 August (*NYT*, 5 August 1977, p. 7) and signed by the President four days later. The Bill revised clean-up rules for nearly all sources of air pollution, and gave cities with dirty air up to ten further years to meet some standards, while offering new protection to some areas with clean air (*NYT*, 5 August 1977, p. 9). It contained clauses to delay and weaken control on vehicle exhaust fumes that constituted potential health hazards. It also had clauses that weakened existing rules that barred activity resulting in deterioration in pristine air in national parks, national forests, and other protected areas. By September 1979, however, the Interior Department was to propose that forty-seven primitive areas and national monuments around the country should be covered by the 1977 Clean Air Act, to protect them from uncontrolled air pollution (*NYT*, 8 September 1979, p. 10).

Economy or environment?

By late 1977 politicians in the US were well aware of the precarious links between environmental policies and economic recovery. On the one hand, economic growth would both require and promote increased energy production, and this would increase air pollution. The Energy Research and Development Administration concluded in September that US air pollution would get worse over the next twenty years because of increased energy production, *not* (as critics had argued) because of increased use of coal (as proposed in Carter's energy programme) (*NYT*, 29 September 1977, p. 13). In turn, air pollution would cause much damage, hence have high costs. The Midwest Research Institute estimated, also in September, that air pollution was costing residents in large US cities *at least* $6 billion each year in damage in property and health (*NYT*, 4 September 1977, p. 22). Damage in New York City was highest at around $1 billion in 1970.

The reverse side of the coin was that vast sums would have to be spent on pollution-control equipment, which would raise prices and reduce competitive abilities in some industries, especially those with large energy requirements. It might, in serious cases, lead to plant closure and unemployment. Already the US steel industry was facing serious difficulties when faced with the need to spend hundreds of millions of dollars on such equipment (*NYT*, 5 September 1977, p. 16). When 250 steelworkers who had been laid off work demonstrated outside the White House in September 1977 (*NYT*, 24 September 1977, p.27), the President noted that foreign industry (particularly in Japan and West Germany) had similar pollution-control constraints, and the root problem was linked to the depressed world economy (*NYT*, 30 September 1977, p. 19).

The compromising attitude of the administration to environmental matters when faced with economic problems came to a head in February 1979, when the White House was charged with intervening in the normal regulatory process by frustrating EPA efforts to carry out the nation's anti-pollution laws (*NYT*, 27 February 1979, p. 7). The President was quick to point out that he had no intention of undermining these laws through his policies to fight inflation and reduce the burden of federal regulation (*NYT*, 28 February 1979, p. 15). Six months later he pledged to Congress that efforts to end the energy crisis would not affect his administration's commitment to clean air (*NYT*, 3 August 1979, p. 24).

Acid rain drops

By early 1978 the focus of the US air pollution debate was to switch, when concern rose over acid rain. Granted that the government was taking air pollution seriously – one-third of the $40.6 billion spent on the environment in 1977 went to reduce air pollution (*NYT*, 1 March 1978, p. 14) – progress in combating the problem was being seen as too little, too late, and too slow.

Alarm rose further when surveys of fishless acid lakes in the Adirondack Mountains, first published in April 1978 (*NYT*, 29 April 1978, p. 25), indicated the likelihood that the damage was caused by contaminated rain and snow originating in coal- and oil-fired power stations and industrial plant in the Ohio Valley and across the border in eastern Canada. Scientific interest in acidification was by now well advanced; research was even under way to develop trout that could survive in the Adirondacks' acid lakes (*NYT*, 12 September 1978, p. 12).

Political interest was also on the increase. In August 1979 New York state Environmental Conservation Commissioner Robert Flacke expressed concern over fish and water quality in 400 of the Adirondack's major lakes (*NYT*, 18 August 1979, p. 23). In the same month the Carter administration launched a ten-year programme of research (managed by the Acid Rain Co-ordination Committee) into acid rain and its possible effects (*NYT*, 12 August 1979, p. 8).

Scientists, environmentalists, and some politicians looked towards the EPA to take a lead in 'sorting out the mess' of acid rain in the United States. It duly obliged in September 1978 (*NYT*, 12 September 1978, p. 12) by calling on electricity utilities to install scrubbers (at a cost of around $10 billion) to filter out SO_2 flue gases, the output of which had increased a lot with the planned massive shift to coal burning. The National Commission on Air Quality reported a year later that acid rain was getting worse and more widespread in the United States (*NYT*, 7 October 1979, p. 61). Most witnesses pointed the finger at coal burning in industry and vehicle exhaust emissions.

Politics, economics, and acid rain

The scene was now set for a prolonged debate on acid rain in the United States.

Some of the critical early moves were made in New York state, where over 7,000 acres of pristine habitats in the Adirondacks had already been seriously affected by acidification. In late January 1980, Governor Carey announced that a group of Northeast states were seeking federal action against acid rain blowing in from the south and west (*NYT*, 28 January 1980, p. 2). Days later, the state's Environmental Conservation Commissioner Robert Flacke telegrammed the EPA to urge immediate action. He proposed tight federal emission standards for SO_2 and NO_x, and mandatory coal washing (*NYT*, 1 February 1980, p. 2). The call did *not* land on deaf ears, because in March EPA Administrator Douglas Costle told a House Commerce Sub-committee (on Oversight and Investigations) that the government should do *something* about acid rain (*NYT*, 2 March 1980, p. 51). But he also noted that deciding what action to take and introducing the laws necessary to back it up could take *at least* ten years.

Not everyone shared the EPA view that acid rain was becoming a serious problem. Neither did everyone agree with the EPA that coal burning was a major cause of the problem. For example, the President's Coal Commission argued that with current air pollution control technologies, the nation could burn two to three times the tonnage of coal that it already was burning without appreciably adding to the air pollution problem, which would, in any case, increase as demand for electricity rose (*NYT*, 3 March 1980, p. 10).

The opposite view was taken by energy officials (*NYT*, 7 March 1980, p. 1), who feared that Carter's ambitious $10 billion programme to induce utilities to burn more domestic coal would increase air pollution and further aggravate the acid rain problem. Douglas Costle of the EPA warned the Senate Environmental Pollution Sub-committee (*NYT*, 20 March 1980, p. 18) that the programme would only increase the acid rain problem (by 10–15 per cent, he announced later – *NYT*, 22 April 1980, p. 5). The announcement that a mere $400,000 – of the $3.6 billion set aside to help the utilities with the cost of conversion from oil to coal – would be available to reduce acid rain (*NYT*, 9 March 1980, p. 39), did little to alleviate such fears. Calls were made to make available much larger sums (*billions* of dollars) to cover the cost of installing emission-control equipment (*NYT*, 12 March 1980, p. 26).

The EPA, now drawn into the very heart of the acid rain debate, sponsored a two-day conference on acid rain in April 1980 (*NYT*, 10 April 1980, p. 22). Although it was designed mainly to focus on the scientific nature of the problem, the meeting gave a long-awaited opportunity for politicians to air their grievances as well. A coalition of states (mainly in the East) claimed that they were being unfairly treated by other states that burned 'dirty' coal in power plants, and they tabled the framework for a 'national policy' to deal with acid rain. Those accused were not convinced, and several Midwest states (which would face massive expenditure if a national clean-up campaign were launched) insisted that proof of cause and effect was not yet established. Eager to provide a platform of reliable understanding on which to base its decisions, the EPA launched a $2 million study of acid rain in July 1980 (*NYT*, 18 July 1980, p. 6).

By mid-1980 the trans-frontier character of acid rain in North America was well established, even if all parties did not agree about the magnitude or precise causes of the problem. Canada insisted that the US was exporting vast amounts of SO_2 and NO_x (thus of acid rain) on to Canadian lakes, soils, and forests. Most of this, it argued, came from the main industrial centres (particularly from coal-burning power plants) in Eastern,

Northern and the Midwest USA. The United States, on the other hand, pleaded 'not guilty' and argued, instead, that most of Canada's acid rain problem was essentially 'home grown'. Diplomatic exchanges between the two countries had grown increasingly strong and at times acrimonious.

A conciliatory gesture of goodwill came in August 1980, when US Secretary of State Edmund Muskie and Canadian officials signed a document committing both countries to work towards a treaty controlling the spread of acid rain and other pollutants (*NYT*, 7 August 1980, p. 17). Canada's Minister of Environment John Roberts estimated that the cost of reducing acid rain and other pollutants in the Northeast USA and eastern Canada would be in the region of $400–500 million over the next twenty years.

REAGAN – THE EARLY YEARS

As soon as the results of the 1980 elections in the US were known, Canadian officials were keen to express their concerns over acid rain to President-elect Ronald Reagan (*NYT*, 8 November 1980, p. 24). Their main worry was that – in his drive to stimulate industry and solve the energy crisis – he would give low priority to the environment. They had read his intentions right, because Reagan's special group on the environment believed that his administration *would* probably seek to modify or disband some anti-pollution regulations that had no economic or scientific justification, beginning with the Clean Air Act (which was coming up for review in 1981 in any case) (*NYT*, 14 November 1980, p. 18). This was seen as a swing away from the environmentalism of the 1970s, made necessary by the state of the nation's economy (*NYT*, 20 November 1980, p. 1).

Not *all* politicians accepted that this was the right way forward. For example, Walter Cronkite argued that if the cost of meeting pollution standards put US companies at a disadvantage when competing with foreign companies, the products from any country that did *not* enforce pollution standards as strict as those in the US should be barred from US markets (*NYT*, 14 October 1980, p. 4).

The Reagan administration's early days in office saw an upsurge of interest in acid rain in North America. Not only was scientific understanding of the scale and character of the problem emerging fast, but also political awareness of the implications of the issue (both at home and abroad, especially with Canada) was sharpening. The issue hit the headlines on 25 January 1981 when the Interagency Task Force on Acid Precipitation announced that acid rain and snow had become even *more* acidic, and were falling over larger parts of the Northeast USA and eastern Canada, over the last three decades (*despite* the Clean Air Act of 1970 and the 1977 amendment) (*NYT*, 25 January 1980, p. 30). Average rainfall in the area was ten times more acidic than the norm, but some rain and snow was up to 1,000 times more acidic, the report said. By April 1981 Douglas

Costle (now a former EPA official) was able to testify to Senate that there *was* enough evidence to show that acid rain was having harmful effects in the Northeast USA and Canada (*NYT*, 10 April 1981, p. 10). Other reports throughout the year gave an indication of the scale of the problem: acid rain damage to materials, forests, lakes, and agriculture in the US was estimated to be in the region of $5 billion each year (*NYT*, 26 April 1981, p. 1), and the problem was most acute in the fifteen states east of the Mississippi, which were regarded as extremely vulnerable to the effects of acid rain (*NYT*, 6 October 1981, p. 20).

Alongside the better scientific awareness of the problem, there was clearer vision of the possible benefits of clean-up programmes. A 1981 EPA study estimated that stricter emission controls on electric power plants could prevent up to 54,000 pollution-related deaths and avert billions of dollars in crop losses in the Ohio River Valley (which has the nation's worst sulphur air pollution) by the year 2000 (*NYT*, 2 March 1981, p. 12). The government could no longer ignore the acid rain issue, and in May 1981 the Council on Environmental Quality (CEQ) was assigned the task of co-ordinating the Reagan administration's policy on acid rain (*NYT*, 24 May 1981, p. 22).

Inevitably, concern was rising in New York state, right in the heart of the most polluted area (see Figure 2.2). By now there was much scientific study and monitoring of acid rain and its effects in the Adirondack Mountains, and startling findings were starting to emerge. Acid damage to lakes and ponds in the Adirondacks was found to be much more widespread than previously thought (*NYT*, 19 December 1980, p. 1). Moreover, the sudden acid surges (see chapter 2) when snow melts in spring were found to be more harmful to freshwater fish than normal levels of acid rain (*NYT*, 13 January 1981, p. 5): 40 per cent of the streams in the Adirondacks had critical levels of acidity during the 1980 snowmelt period (*NYT*, 3 November 1981, p. 5). The state called for immediate federal action to bring acid rain under control, without which 'we can look forward to continued destruction of our Adirondack Mountains', concluded Commissioner Robert Flacke (*NYT*, 19 December 1980, p. 1).

Also inevitable was the growing tide of opinion within the US in favour of tackling the acid rain issue in a *realistic* way. In February 1981 the National Commission on Air Quality recommended that acid rain should be dealt with without further delay (*NYT*, 28 February 1981, p. 28), and in the same month the National Governors' Association (representing forty-eight states) declared its view that the best way would be via state-to-state arrangements rather than federal controls (*NYT*, 24 February 1981, p. 10).

Equally inevitably, political tensions were starting to appear between Eastern states (receivers of acid rain) and Midwest states (donors of SO_2). In January 1981, New York Attorney-General Robert Abrams filed a petition with the EPA in an attempt to block an application from several power plants in the Midwest to allow them to increase SO_2 emissions from their coal-burning power plants (*NYT*, 28 January 1981, p. 7). To grant such permission, he stressed, would be to grant them a licence further to aggravate the acid rain problem in New York state. Irritation with Midwest states was high. One letter to the *New York Times* on 17 June 1981 (p. 30) argued that, if the Midwest wanted the benefits of coal-fired power, then it should use short smokestacks so that serious environmental damage (from dry deposition) was kept close to home and *not* exported (as acid rain) to the Northeast. The East–West debate was kept raging when the National Academy of Sciences announced that it had evidence that acid rain in the Northeast was caused primarily by the region's *own* coal-fired electricity plant, rather than those in the Midwest (as widely argued) (*NYT*, 20 September 1981, p. 24).

Inevitable, too, was the mounting Canadian concern over US exports of acid rain. When President Reagan addressed the Canadian parliament on 12 March 1981, he pledged strong friendship between the two countries but offered *no* concessions on trans-border trade in acid rain (*NYT*, 12 March 1981, p. 12). Canada's Environment Minister John Roberts later noted that – in light of the fact that 70 per cent of the acid rain that fell in Canada was believed to come from the United States – his government was growing 'increasingly impatient with US failure to control acid rain' by failing to meet its commitments to resolve *its* part of the problem (*NYT*, 31 March 1981, p. 17). Acid rain was threatening to turn Canada into a bitter enemy of the United States (*NYT*, 30 April 1981, p. 31), and Canadian Prime Minister Pierre Trudeau raised the issue on 10 July 1981 with President Reagan, who was in Ottawa for the annual bilateral economic summit meeting (*NYT*, 11 July 1981, p. 30). The two countries also agreed to disagree at a House Sub-committee hearing in October 1981 (*NYT*, 7 October 1981, p. 12) when Raymond Robinson (Canada's Assistant Deputy Minister) challenged the view of EPA official Kathleen Bennett that not enough was yet known about acid rain to allow realistic decisions to be made by the two governments about remedial action. That same month the Canadian parliament conceded that a twelve-month study of acid rain in Canada had established that about half of the country's acid rain was 'home grown', but insisted that this still did *not* absolve the USA from responsibility for causing most of the observed damage to Canadian lakes and farmland (*NYT*, 9 October 1981, p. 9).

Changing perspectives (1982)

If Reagan's early days were marked by growing stress over many aspects of the acid rain problem, the following year (1982) was to see marked shifts in perspectives and growing political sensitivity.

Tangible evidence of damage from acidification was becoming even more abundant. There was now little doubt that acid rain had seriously altered the chemical balance of thousands of lakes and streams in the United States (*NYT*, 26 March 1982, p. 12), and of several thousand lakes in Ontario (*NYT*, 19 October 1982, p. 12). Warnings were also given that acid damage was likely to spread far and fast – fears were expressed for the survival of Atlantic salmon in many rivers in Northeast USA and Canada (*NYT*, 23 February 1982, p. 22). In November 1982 the Izaak Walton League reported that a much larger portion of the US was susceptible to damage from acid rain than was previously realized (*NYT*, 9 November 1982, p. 25). Highly acid rain and snow had already been found in some parts of Colorado, and wildlife in some of the state's lakes appeared to be threatened (*NYT*, 11 August 1981, p. 11).

Public concern over acid rain was also on the increase, and environmental groups began their efforts to mould public opinion in favour of emission control and to warn the public of the likely consequences of failing to address the problem without further delay. Keen to attract widespread media coverage, five Greenpeace activists scaled smokestacks in Ohio. Indiana, and Arizona in February 1982 to protest against acid rain (*NYT*, 11 February 1982, p. 31; 13 February 1982, p. 23). Their lofty vigil was effective but short-lived; two days later they were arrested and charged with trespass. Canada joined the media bandwagon in November of that year, when diplomats and employees at its Washington embassy sported umbrellas that called attention to the acid rain problem (*NYT*, 6 November 1982, p. 9).

Acid policy and politics also continued at state and national levels. By now deficiencies in the Clean Air Act (1970, with 1977 amendments) were becoming glaringly obvious, especially given the new-found significance of acid rain in the nation's consciousness. Calls to address the problems of acid rain in revising the Act were made by – amongst others – the National Audubon Society (*NYT*, 25 March 1982, p. 30) and the American Lung Association (*NYT*, 28 March 1982, p. 16). In July 1982 the Senate Environmental Committee drafted amendments that would require power plants and other industries that burn coal and other fossil fuels to reduce their SO_2 emissions by about half (8 million tonnes) over the next twelve years, to help solve the problem of acid rain (*NYT*, 23 July 1982, p. 16).

However, the nation was *not* in full agreement about the need for radical changes. For example, the EPA – which under Carter had taken a very positive line on acid rain – was clearly changing its colours. Early in 1982 it did an about-turn, and refused to endorse its earlier conclusion that the scientific evidence of damage from acid rain was so persuasive that far-reaching remedial measures were required without delay (*NYT*, 9 February 1982, p. 3). To add to the confusion, the Senate Environment Committee, at its May sitting, heard many conflicting opinions from scientists on whether enough really was known about acid rain to justify action without further delay (*NYT*, 27 May 1982, p. 13).

The need for more scientific investigations of the causes and controls of acid rain was widely agreed, and it was an argument frequently rehearsed by the Reagan administration. The rhetoric looked a bit hollow, and the administration's commitment to *do* anything to enhance scientific understanding came into question when, in December 1982, it sharply reduced funding for its major research programme on air pollution by electricity utilities (*NYT*, 2 December 1982, p. 17). This followed the first blow, six months earlier, of cutting off funds to the National Academy of Sciences for acid rain research (*NYT*, 8 June 1982, p. 1) after it had urged the EPA to consider restrictions on sulphur emissions in order to reduce acid rain. This cynical move reflected the administration's belief that the Academy had a 'biased' view on acid rain, but it immediately attracted criticisms that it was allowing politics to interfere with legitimate and significant scientific enquiry (*NYT*, 8 June 1982, p. 1). Tension between science and politics was soon to become a hallmark of the acid rain debate in the United States. Similar political doggedness was at the root of the Reagan administration's stubborn withholding of approval of the OECD report on the global environment, because of its objections to proposals for governmental action to curb acid rain (*NYT*, 28 March 1982, p. 4).

Tension between Canada and the United States was also high, and it was aired in public. Canada's Deputy Environmental Minister (J. Seaborn) openly criticized US policy on acid rain at a UN conference on the global environment, held in Nairobi in May 1982 (*NYT*, 14 May 1982, p. 5). During the same month the Canadian government increased its lobbying efforts in the US Congress to ensure that the Clean Air Act be extended to include provisions limiting acid rain (*NYT*, 27 May 1982, p. 17). Later that year the National Wildlife Federation accepted a donation of $10,000 from its Canadian counterpart to support its fight against acid rain. This was taken – by CEQ Chairman Alan Hill – as a sign that the US Federation had a 'biased' point of view (*NYT*, 2 September 1982, p. 8). But not all contact between Canada and the US was so contentious – in July 1982 New York state and Quebec signed a bilateral agreement to co-ordinate efforts to combat acid rain (*NYT*, 27 July 1982, p. 3).

Political initiatives were also being taken in New York state. In March 1982 Governor Carey proposed a quick 'minimum program' to reduce acid rain damage, involving strict limits on SO_2 emissions from coal-fired power stations and coal washing for coal with a sulphur content higher than 3 per cent (*NYT*, 10 March 1982, p. 2). The state also pledged to take the lead in pressing for federal legislation and inter-state mediation panels to help control the spread of acid rain (*NYT*, 11 March 1982, p. 5).

The powerful US coal lobby also had a high profile in the debate during 1982. There were claims of a Senate bias against coal and that anti-coal bills would be premature. Politicians in 'coal states' (like Indiana) argued that air pollution legislation would do serious harm to the coal industry (*NYT*, 25 July 1982, p. 21). At the same time support was given in Congress to measures that would require that all coal be ground and washed before leaving the mines (*NYT*, 13 May 1982, p. 26) ('coal cleaning' – see chapter 7). Proposals to allow conversion of three power stations in New York state from oil to coal were strongly attacked by Dr Michael Oppenheimer of the Environmental Defense Fund (EDF) (*NYT*,

1 December 1982, p. 2), on the grounds that they would further aggravate already serious problems of acidification in the Adirondack Mountains.

MOUNTING INDECISION (1983)

By 1983 acid rain was being heralded as an unwanted problem in many countries, especially in Europe (chapter 8). Forest decline on a massive scale was becoming apparent in the Black Forest of south-west Germany (*NYT*, 6 November 1983, p. 1), and acid rain was suspected as the main cause. Growing concern over possible damage from acid rain prompted a five-year scientific study into its causes and effects in Scandinavia (*NYT*, 6 September 1983, p. 5), funded by Britain (see chapter 10), and involving scientists in Britain, Norway, and Sweden. In North America, acid rain threatened to sour even further relations between Eastern and Midwestern states, and between the US and Canada.

There was growing concern in New Jersey and other Northeastern states about widespread damage by acid rain to lakes and ponds (*NYT*, 22 May 1983, p. 34), and by March 1983 over 100 communities in New Hampshire had passed resolutions asking the federal government to act on acid rain by requiring reductions in industrial emissions of SO_2 in the Middle West (*NYT*, 10 March 1983, p. 20). Whilst new detection methods had by now been developed, which could demonstrate that chemical emissions from the Midwest fell as acid rain in the East (*NYT*, 29 August 1983, p. 12), some politicians continued to insist that there was *no* conclusive evidence that Midwest industry was to blame for acid rain problems in Eastern states (*NYT*, 14 December 1983, p. 14).

Whilst such critical issues remained open to debate, little substantial progress could be made in dealing with the growing problems of acidification in the East. Responsibility was a key question, but equally significant was the question of *who* would pay (and *how*) for clean-up programmes. In December 1983, twenty-six members of Congress from Western states complained to the President that proposals to reduce acid rain nationally would unfairly tax their region: states in the East would benefit, they would pay the cost (*NYT*, 3 December 1983, p. 17). Two months earlier, governors attending the Midwestern Governors' Conference rejected a statement calling for a national tax on SO_2 emissions, and voted instead to call for an accelerated federal programme to cut emissions (*NYT*, 11 October 1983, p. 22). They approved a policy statement calling for a freezing of SO_2 emissions at 1980 levels until a national plan could be enacted.

Politicians in Eastern states were by now frustrated that nothing seemed to be being done about acid rain problems in their area, and were willing to make quite sizeable concessions to show goodwill and determination to tackle the problem. They sought conciliatory but *realistic* ways forward, rather than simply pointing the finger at other states. For example, in December 1983 a coalition of six Northeastern governors discussed ways of reducing acid rain (*NYT*, 6 December 1983, p. 1; 11 December 1983, p. 9), and proposed a $5 billion annual plan, based on cost sharing (between polluters and polluted). This involved, amongst other things, establishing a national $2 billion trust fund to help electricity-generating companies to pay for the installation of SO_2 control equipment.

Charity begins at home, and initiatives to cut SO_2 emissions began in the East. In September 1983, New York state launched a programme designed to restrain SO_2 emissions within the state, which it hoped would point the way for other states, Congress, and the EPA in their search for viable solutions to the acid rain problem (*NYT*, 24 September 1983, p. 23). The state also formulated a three-year, $4.2 million study of water quality and fish in 1,200 Adirondack lakes, to be funded mainly ($3.52 million) by state public utilities and administered by a newly created, non-profit Adirondack Lakes Survey Corporation (*NYT*, 20 November 1983, p. 51).

Canadian concern

Relations between the US and Canada were particularly troubled by the acid rain issue, and in February 1983 Canadian Environment Minister John Roberts called for a 50 per cent reduction in SO_2 emissions in the US (*NYT*, 22 February 1983, p. 5).

Relations were to get much more frosty, when in March the US Justice Department ordered that two Canadian films on acid rain be labelled foreign 'political propaganda' (*NYT*, 3 March 1983, p. 27). Even within the US, the claim was made that this blatant act of cynical censorship showed the Reagan administration's fear of open debate and ready availability of information. The Attorney-General (William Smith) was quick to point out that the 'eccentric' decision was taken *not* by him but by 'zealous subordinates' acting alone (*NYT*, 6 March 1983, p. 18). The 'propaganda' films reaped the unwanted benefit of notoriety, courtesy of the Justice Department rating, and they were widely shown (e.g. on New Jersey Network TV later in March – *NYT*, 27 March 1983, p. 10) and eagerly debated.

Calls from Canadian scientists and politicians for the US to curb SO_2 emissions continued through the year (*NYT*, 28 July 1983, p. 2). Some US Congressmen – ever eager to discredit critics of their non-committal stance on acid rain – shouted 'foul', and dismissed this strong lobby as foreign interference in American politics, which was 'brazen' and could lead to 'loss of some kind of sovereignty and separate identity' (*NYT*, 21 August 1982, p. 15). The Canadian pressure was maintained, regardless of this US accusation. Canada by now had a new environment minister – Charles Caccia – who reported an encouraging change in US attitudes to acid rain when he met the new EPA head William Ruckelshaus in September 1983 (*NYT*, 9 September 1983, p. 9). Ruckelshaus promised to prepare, within one month, a new policy for dealing with acid rain; as we shall see, it was to be a hollow promise. When the promised plan failed to materialize, Caccia announced that the Canadians were growing increasingly impatient with US inertia over acid rain (*NYT*, 17 October 1983, p. 1).

September also saw the first large-scale monitoring of long-distance polluted plume movements between the US and Canada, using a network of small planes and ground stations to track the eastward passage of SO_2 and NO_x (*NYT*, 22 September 1983, p. 14).

Pressure rises

By late 1983 relations between the US and Canada were strained because of acid rain, and inter-state dialogue within the US was getting fraught if not plain acrimonious. It was clear that Congress could not avoid the issue any longer, and that it should seek to tackle acid rain by reducing emissions of SO_2 and NO_x (*NYT*, 1 November 1983, p. 27). There was *some* support for a proposal from Senator Robert Stafford that would order a specified cut in SO_2 emissions but let the power companies decide how to reach the target (*NYT*, 3 September 1983, p. 22).

The Reagan administration was faced with powerful industrial lobbies for and against pollution control. Clearly in a matter like control of acid rain, some industries stood to gain and others to lose. Amongst the winners would be manufacturers of the pollution-control equipment, and the Industrial Gas Cleaning Institute saw the prospect of federal legislation as the 'only ray of sunshine' for new businesses in the depressed scrubber manufacturing industry (*NYT*, 7 November 1983, p. 1). Amongst the losers would be the electricity utilities, which would have to meet most of the cost of installing control equipment. It was little surprise, therefore, when the Edison Electric Association took the initiative in founding a body, which it invited other trade organizations and companies to join – the Alliance for Balanced Environmental Solutions – that would seek to educate the general public about acid rain, and seek to combat legislation aimed at requiring utilities to install costly control equipment in power stations (*NYT*, 3 February 1983, p. 6). In June 1983, the Edison Electric Institute reported on a survey of twenty-four utility companies, which found that proposals to control acid rain that were then before Senate would require a 14 per cent increase in electricity rates in some parts of the US (*NYT*, 29 June 1983, p. 13).

In June 1983 the EPA got a new head, William Ruckelshaus. He had earlier told the Senate committee considering his nomination that he favoured more study before acting to control acid rain, and he favoured changes in the Clean Air Act (*NYT*, 6 May 1983, p. 15). He took up the challenge of acid rain with enthusiasm, and within a month had assembled a panel of fifteen scientists to deal with the issue (*NYT*, 21 July 1983, p. 10). At the beginning of September he announced his intention to recommend a national plan to deal with acid rain, giving the President four or five options to choose from (*NYT*, 1 September 1983, p. 14). The plan would be prepared within a month, and cost an estimated $1.5–2.5 billion.

Ruckelshaus devoted considerable attention to the formulation of the plan, but his eagerness won him the antagonism of the powerful Energy Department and the Office of Management and Budget – who saw the plan as premature and too costly (*NYT*, 27 September 1983, p. 1). By late October, instead of unveiling the eagerly awaited plan – as intended – the hapless Ruckelshaus was forced to announce an indefinite delay in making recommendations to the President for controlling acid rain because of opposition from high-level administration officials (*NYT*, 23 October 1983, p. 27). This failure to 'produce the goods', especially after such public promises, attracted much criticism (*NYT*, 20 November 1983, p. 4).

If 1983 was to be a year of pendulum swings in US attitudes to the control of acid rain, this reflected the fickleness of political arguments more than the rationality of scientific understanding. A wave of reports on the science of acid rain appeared during the year, all pointing to the seriousness of the problem and most to the need to take action without delay.

In June, the Interagency Task Force reported that man-made air pollution – SO_2 and NO_x from power stations, industry, and vehicles – was the most probable cause of the acid rain that was disturbing freshwater rivers and lakes in the Northeast (*NYT*, 9 June 1983, p. 1). It did, however, stress that not enough was known about the links between emissions, deposition, and damage. It was very much a detached scientific review, which did *not* recommend any remedies. But the factual revelations in the report were eagerly seized upon by Michael Oppenheimer, who concluded that the major question now before Congress should be where to start the clean up, and who should pay for it (*NYT*, 15 June 1983, p. 27). In the same month, a panel of scientists reported to the White House Science Office (the President's science adviser) its conclusion that – despite continued uncertainty – action should be taken without delay to curb acid rain (*NYT*, 28 June 1983, p. 14). The administration's position remained that more research was needed.

In late June, the National Academy of Sciences (NAS) joined what was then being described as the 'increasingly urgent national debate over acid rain' (*NYT*, 29 June 1983, p. 1), with the publication of a report that argued that the links between emissions of SO_2 and NO_x from tall smokestacks and acid rain were now well proven. The report established that the amount of acid rain was roughly proportional to the amount of SO_2 and NO_x emitted by polluters (i.e. the link was linear – see chapter 3). It concluded that acid rain *could* be curbed by reducing SO_2 emissions from coal-burning power stations and factories in the Eastern US, especially in the Ohio Valley (*NYT*, 30 June 1983, p. 16; 1 July 1983, p. 22).

The NAS report betrayed the lack of substance to the Reagan administration's over-used argument that regulatory programmes must await better understanding of the causes and effects of acid rain. In August, the President's Council on Environmental Quality announced its intention to create a panel of specialists to review the federal research programme on acid rain (*NYT*, 21 August 1983, p. 15). Once again, the US people were to witness political indifference to reliable scientific understanding, and continued 'fence-sitting' by an administration that was fast running out of excuses for reluctance to curb acid rain by reducing oxide emissions.

RESISTANCE AND ENTRENCHMENT (1984)

The spread of acid rain damage across Europe during 1984 (described in chapter 8) was matched by mounting evidence of damage in North America. The acid alarm was raised when scientists reported evidence of serious forest decline – similar to that being observed in Europe (chapter 5) – throughout the Northeast USA. Death of softwood species in the higher parts of the south-eastern Appalachians, and spruce and fir trees on Mount Mitchell (North Carolina) was attributed to acid rain (*NYT*, 26 February 1984, p. 1). Tree damage was particularly acute along the heavily polluted Ohio Valley (*NYT*, 15 April 1984, p. 19). Acidic lakes in upstate New York and northern New England also appeared to be getting worse (*NYT*, 6 March 1984, p. 10), even though sulphate levels flowing into them were believed to have fallen.

Alarm was also caused by the publication in April of a report by the National Wildlife Federation (NWF) stating that acid rain was a problem throughout much of the nation, not just in the Northeast as previously thought. In the same month an Environmental Defense Fund report (*NYT*, 23 April 1984, p. 12) concluded that damage to lakes and streams in the US was likely to get *much* worse if acid rain continued to fall at current levels (*NYT*, 27 April 1984, p. 17).

Two months later, a report from the National Clean Air Coalition was to conclude that rainfall over much of the thirteen Southern states in the US was ten to twenty times more acidic than normal rain, and that the South faced the prospect of acid rain damage similar to that in Canada and the Northeast (*NYT*, 8 June 1984, p. 15). The area under threat was later to be extended to include Wyoming after the state's Environmental Quality Council was warned by US Forest Service officials that rainfall ten times more acidic than normal was being measured there (*NYT*, 18 November 1984, p. 33). Also in November, lakes and streams in the Rocky Mountains were being threatened with acidification believed to be caused by SO_2 emissions from two large copper smelters downwind in Arizona (*NYT*, 29 November 1984, p. 27).

Presidential prerogatives

Both the NWF and the EDF reports concluded that current scientific evidence on acid rain *was* adequate to control the problem, and that the Reagan administration and Congress should act quickly to solve the growing nationwide crisis. Inevitably, given the widespread publicity and broadening base of public interest, the acid rain issue became an important ingredient in the 1984 presidential campaign.

The campaign trail was littered with aspiring candidates and their supporters – including Reverend Jesse Jackson, Walter Mondale, and Senators John Glenn and Gary Hart (*NYT*, 7 January 1984, p. 8) – eager to air their views on acid rain. Democratic presidential hopeful John Glenn (of moon-walk fame) promised to bring in a bill to cut SO_2 emissions from coal-burning power plants (*NYT*, 8 January 1984, p. 23). Walter Mondale, a contestant in the same battle, dismissed Glenn for his unpromising record of voting in Senate on environmental issues and the incumbent Reagan for his inactivity on acid rain; with an eye fixed firmly on the popular vote, he offered his own plan for a 50 per cent cut in SO_2 emissions from coal-burning power plants (*NYT*, 9 January 1984, p. 7). Republican Geraldine Ferraro also pledged to cut SO_2 emissions by half (*NYT*, 16 August 1984, p. 11).

As it turned out, this lottery of offerings on SO_2 reduction was to be a side-show to the main election issues, and, as the history books show, Ronald Reagan was re-elected for a second term in the White House. Reagan, a showman to the end, with a keen sense of awareness of when to throw concessionary crumbs to the snapping electorate, announced in mid-January that he was 'moving towards' proposing additional steps to curb acid rain in a programme that would cost substantially less than the one previously advocated by the EPA (*NYT*, 18 January 1984, p. 21). In one blow he bought off much public opposition to his inactivity *and* even looked like emerging as the good guy with a cheap and effective solution to a problem that simply would not go away.

However, exactly what these intended steps were – if, indeed, they ever seriously existed at all other than as concessionary rhetoric – was never to be revealed. At a meeting with the President only days after his public pledge, a group of Northeastern and Midwestern governors were given no commitment by him that he would take immediate action to curb acid rain (*NYT*, 21 January 1984, p. 14). Whilst the President *did* double the 1985 budget for research on acid rain from $27 million to $55 million, he did *not* recommend any programme to control the air pollution that produced it (*NYT*, 25 January 1984, p. 14).

His unwillingness to act was strongly attacked by members of Congress, the Canadian government, and environmentalists in the USA and Canada. EPA head William Ruckelshaus commented that the President made a significant change in policy by recognizing that the problem existed, even if he did not believe that there were enough data on problems to warrant a costly programme to control acid rain *at source* (*NYT*, 27 January 1984, p. 9). He did, however, concede that – whilst this did *not* serve as a basis for any decisions – the administration had carefully weighed the relative electoral strengths of the polluted and the polluting states (*NYT*, 6 February 1984, p. 11).

The President's commitment to deal with acid rain was questioned further after the White House was accused (in August 1984) of suppressing for more than three months a report by independent scientists that said smokestack pollutants cause acid rain (*NYT*, 18 August 1984, p. 10). Despite strong denials by the White House and William Nierenberg (chief author of the report), the impression was given that the administration wanted to 'keep the lid' firmly on the acid rain debate at this time. The report, commissioned by the President's Office of Science and Technology, was eventually published in September (*NYT*, 5 September 1984, p. 7). It recommended an immediate start on cost-effective steps to control acid rain at source, by reducing emissions of SO_2 and NO_x – an approach diametrical to that of the President, who remained committed to study rather than attack the problem (*NYT*, 27 April 1984, p. 27). He was criticized repeatedly for failing to accept that government studies already showed the need for immediate action against acid rain (*NYT*, 26 December 1984, p. 30).

The President's ultra-cautious attitude partly reflected the strength of feelings against legislation to control oxide emissions that persisted in some sectors of the US economy. The National Coal Association, for example, remained staunchly opposed to such legislation – it had, after all, a lot at stake. It went one step further than shouting 'not guilty'; it claimed 'no case to answer'. The Association argued, for example, that the link between power station emissions and acid rain had *not* been conclusively proved, that the problem was neither *widespread* nor a threat to *health*, that it did *not* appear to be getting worse, and that the problem could be solved by more cost-effective and well-tried methods (*NYT*, 28 October 1984, p. 2).

Whilst all this was going on, problems of acidification continued to get worse, particularly in the Eastern states. Faced with rising and spreading damage, feelings of frustration were growing stronger by the day. In January five states (Maine, Minnesota, Rhode Island, Vermont, and New York) informed the EPA of their intention to sue it over its failure to order reductions in SO_2 emissions that cause acid rain (*NYT*, 19 January 1984, p. 1). By late March, Connecticut had joined the five renegades, and a suit was filed against the federal government accusing the EPA of ignoring sections of the Clean Air Act under which states in the Middle West would be forced to reduce their SO_2 emissions (*NYT*, 21 March 1984, p. 13). Petitions against the EPA for failing to curb acid rain stemming from the Midwest had already been filed in 1980 and 1981 by New York, Maine, and Pennsylvania, and these were rebuffed by the Agency in August 1984 (*NYT*, 30 August 1984, p. 1) and formally rejected by it in December (*NYT*, 6 December 1984, p. 13). The EPA rejection of the petitions was widely deplored (*NYT*, 2 September 1984, p. 14), and seven Northeastern states appealed against it (*NYT*, 7 December 1984, p. 22).

The East–West dialogue also surfaced at the National Governors' Association meeting in Washington, in February (*NYT*, 26 February 1984, p. 31). The NGA committee gave thought to calling for a reduction of SO_2

emissions by 10 million tonnes over the next ten years, but some states (like Illinois) strongly opposed any plans to specify required reductions or tax offenders.

In June 1984 New York state, once again keen to lead the pack, formulated an ambitious plan to reduce acid rain in the state over the next ten years by sharply limiting emissions from smokestacks *within the state* (*NYT*, 19 June 1984, p. 5). The plan required utilities to reduce their SO_2 emissions by 30–50 per cent (*NYT*, 8 June 1984, p. 26). This was seen as a small but necessary first step towards solving the state's acid rain problem. Michael Oppenheimer of the EDF quickly pointed out that the bill was *no* panacea, and certainly no substitute for federal legislation, since the reduced emissions would alone do little to solve acid problems in the state and be of even less benefit to other Eastern states (*NYT*, 7 July 1984, p. 23). But this was the first law in the US designed specifically to reduce acid rain (*NYT*, 30 June 1984, p. 26), and the state's primary aim was to put pressure on Congress to adopt realistic *nationwide* standards, and thus force a reduction in power station emissions in the Midwest.

International tension

International relations were also being seriously affected by the Reagan administration's stubborn refusal to act on acid rain. Canada, claiming to be on the receiving end of the massive US export of SO_2, marshalled international attention on its intransigent neighbour. In February, environment ministers from eight West European countries visited Ottawa, Canada, to discuss the growing menace of acid rain (*NYT*, 7 February 1984, p. 5). The next month saw the birth of the '30 per cent club' (see chapter 8), also in Ottawa (*NYT*, 22 March 1984, p. 9), and the growing isolation of Britain and the United States – who refused to join – in the debate. Both countries took a similar stance at the ECE meeting in Geneva in September (see chapter 8), when many others joined the '30 per cent club' (*NYT*, 29 September 1984, p. 3).

Dealings between the USA and Canada on the acid rain issue became more and more entrenched. In late February, the Canadian government presented a formal diplomatic protest to the US State Department protesting at the administration's insistence on more research (the administration was charged with stalling for time by insisting on further study) and its failure to offer specific plans to reduce acid rain (*NYT*, 8 February 1984, p. 2; 23 February 1984, p. 5). Canadian Ambassador Allan Gotlieb claimed that acid rain was now the most important issue of common interest between the countries: half of Canada's acid rain was believed to come from across the border in the US (*NYT*, 23 February 1984, p. 5), and Canadian officials felt that US efforts to curb it had been minimal (*NYT*, 26 February 1984, p. 18).

Within days, in early March 1984, Canadian federal and provincial environmental officials announced that the Canadian government would proceed on its own – without US endorsement – to cut emissions of SO_2 by 50 per cent by 1994 (*NYT*, 8 March 1984, p. 16). Environment Minister Charles Caccia urged the US government to join this unilateral initiative (*NYT*, 7 April 1984, p. 24), but his plea fell on deaf ears. In June, Canadian Prime Minister Pierre Trudeau proposed, during an economic summit conference in London, England, that the conference commit itself to take strong action to reduce acid rain (*NYT*, 11 June 1984, p. 1). Opposition to the proposal was led by none other than President Reagan.

After the Canadian elections of 1984, the new prime minister, Brian Mulroney, visited President Reagan at the White House on 25 September, but emerged dejected at the lack of progress in persuading him to move swiftly to curb acid rain (*NYT*, 29 September 1984, p. 3). However, Mulroney was clearly *not* going to let the subject drop: US–Canadian diplomatic exchanges were to continue; Canadian frustration was to rise ever further; US refusal to budge on the issue persisted. It had all the ingredients of a long-lived soap opera, only this time it was for real.

There was by now a tide of support from within and beyond the US for *action*, rather than words or more study. Debates had taken place in committees of Senate and the House of Representatives. In March 1984, for example, the Senate Environment and Public Works Committee approved a plan calling for a 10 million tonne reduction in SO_2 emissions over the next ten years (*NYT*, 8 March 1984, p. 6); a major share of the costs would be met by the small number of industrial states that produced most of the pollution. However, any prospect of introducing federal acid rain legislation – perhaps through reauthorizing the Clean Air Act – was lost when the House Energy and Commerce Sub-committee voted against it in May (*NYT*, 3 May 1984, p. 17).

Encouraging signs were none the less coming from the EPA, now apparently taking a more positive attitude to acid rain after its early-Reagan setback. For instance, in November the Agency proposed new, tighter restrictions on the use of tall smokestacks by power plants and factories as a means of complying with the Clean Air Act (*NYT*, 9 November 1984, p. 16). It also proposed the creation of an Office of Acid Deposition, within the Agency, to co-ordinate all research on acid rain (*NYT*, 14 November 1984, p. 24). But a web of federal bureaucracy soon entangled the proposal, which was not taken any further.

PROBLEMS AT HOME AND ABROAD (1985)

By 1985 the key ingredients in the US debate had emerged: much damage in the Northeast states, growing signs of damage in Western states, continuing dialogue between the USA and Canada over the issue of trans-frontier pollution, continuing resistance in the Reagan administration to order sizeable reductions in SO_2 and NO_x emissions.

Science

There was by now abundant evidence that acid rain was developing into a *national* problem in the United States, whereas previously it was seen as a problem largely confined to the Adirondack Mountains in New England and New York. Although SO_2 emissions seemed to be declining in the Northeast, they appeared to be increasing in Southeast states (*NYT*, 29 January 1985, p. 1). In March, the World Resources Institute reported that lakes and forests in mountain areas in the West might be just as vulnerable to acid rain as those further east (*NYT*, 29 March 1985, p. 23), and this extension of susceptibility to the Midwest was confirmed in updated maps published by the EPA in July (*NYT*, 2 July 1985, p. 12). Fears were expressed that air pollution from the region's many smelters was increasing the acidity of high mountain lakes and streams, reducing visibility in some national parks, and perhaps even adding to health problems of residents downwind from the major factories (*NYT*, 30 March 1985, p. 6).

Officials from the National Clean Air Fund and the National Resources Defense Council later added their voice to the debate, in declaring that thousands of lakes and streams in the Midwest (including areas in some of the best-known national parks) were by now highly susceptible to damage from acid rain (*NYT*, 9 August 1985, p. 15). Surveys carried out by the EPA during 1984–5 had found few highly acidic lakes in the East, Southeast and upper Midwest states but a high risk of acidification in these areas (*NYT*, 30 August 1985, p. 16), and in September 1985 the Agency announced its intention (jointly with the National Forest Service) to launch a major border-to-border survey to establish whether there *was* acid rain damage in the West (*NYT*, 18 September 1985, p. 14).

The spread of acid rain across the country convinced the EDF's Michael Oppenheimer that more states would have vested interests in discussions over acid rain legislation (*NYT*, 20 February 1985, p. 23). He believed that would ensure that the matter was taken more seriously and would probably increase the prospects of realistic legislation being introduced without further delay. A further catalyst for action was to be the revelation in a study conducted for the US government by the EPA, Brookhaven National Laboratory, and US Army Corps of Engineers that acid rain was causing around $5 billion of corrosion damage to buildings and structures each year in a seventeen-state region (*NYT*, 18 July 1985, p. 12).

Research on acid rain processes in North America was by now well under way, and important results were beginning to emerge. For example, in August scientists at the Environmental Defense Fund offered the first direct evidence to link changes in pollution emissions hundreds of kilometres upwind to changes in rainfall acidity (*NYT*, 23 August 1985, p. 1). Government experts remained unconvinced about the correlation between emissions and deposition, but many scientists heralded the evidence as a major advance in acid rain research. Remedial measures were also being carefully studied. Scientists from Cornell University evaluated the Swedish technique of liming acidified lakes (see chapter 7) on ten 'dead' lakes in the Adirondack Mountains (*NYT*, 30 April 1985, p. 2). Temporary recovery was possible by this buffering agent, which allowed brook trout – long since absent from the lakes – to be successfully introduced.

Politics

The East–West debate in the United States continued through 1985. Surveys in New Jersey, sponsored by the US Forest Service, were starting to establish how and how much acid rain was affecting forests in the state (*NYT*, 19 May 1985, p. 4), and a detailed statewide study showed that the average acidity of rain, sleet, and snow in the state was at least thirty times higher than the norm (*NYT*, 19 May 1985, p. 1). The same study confirmed that the highest acidities in precipitation over New Jersey were associated with storms coming in from the west and west-northwest – lending substance to claims that air pollution from the Midwest was responsible for acid rain (and damage) in the East.

A certain paradox surrounds the issue, because a 1984 report by the environmental group Inform found that about 40 per cent of New Jersey's electricity was being bought from utilities in the Midwest, from where the emissions originated (*NYT*, 26 May 1985, p. 8); to some extent New Jersey was contributing to its own acidification problem. One solution was offered by Assembly-Woman Maureen Ogden, who in May 1985 proposed state legislation that would prohibit electric companies in the state from buying in power from out-of-state generators that contributed to the state's imported acid rain problem (*NYT*, 26 May 1985, p. 8).

July 1985 was to witness a major turning point in East–West relations within the US, when Federal District Judge Norma Holloway Johnson ordered the EPA to require seven Midwest and border states to reduce their emissions of SO_2 and other pollutants (*NYT*, 28 July 1985, p. 20). The ruling was a direct response to the lawsuit filed by seven Northeastern states in 1984, in an attempt to reduce acid rain in the Northeast and Canada. It was welcomed by New York Governor Mario Cuomo (*NYT*, 29 July 1985, p. 5) and its implications were carefully scrutinized by the EPA (*NYT*, 30 July 1985, p. 11). In December that year the same seven states, along with four environmental groups, filed a suit in New York charging that the EPA had violated the Clean Air Act by failing to obey the requirements to update 1971 standards (*NYT*, 6 December 1984, p. 17).

Canadian concern over acid rain was mounting. Early in the year the Canadian authorities, aware that they should start to put their own house in order, imposed more stringent regulations to cut acid rain (*NYT*, 14 February 1985, p. 15). The plan, announced in February, required a 50 per cent reduction of SO_2 and NO_x emissions in eastern Canada over the next nine years (*NYT*, 7 March 1985, p. 6). This, it was believed, would make a significant contribution to reducing acidification damage in eastern

Canada. Such a gesture of accommodation would also strengthen the negotiating position of Prime Minister Brian Mulroney who was due to meet President Reagan in Quebec City in March. Mulroney's decision to press ahead with pollution control, even without any commitment by the US, was welcomed by the US National Coal Association (*NYT*, 17 February 1985, p. 19).

The urgency of Canada's growing acid rain problem became even more evident at the start of the sugaring season (March), when maple syrup makers found a marked decline in the health of their maple trees (*NYT*, 6 March 1985, p. 14). When even the country's national emblem is threatened by acid rain, then things really are getting serious.

Canadian commitment to tackle the acid rain problem was also evident in December 1985 when Ontario's Environment Minister announced that the four largest industrial sources of SO_2 in the province had been ordered to reduce emissions from 1.9·million tonnes to 665,000 tonnes (a 65 per cent drop) by 1994 (*NYT*, 18 December 1985, p. 12). It was believed that such a step would significantly reduce acid rain across the border in New York state and New England, as well as in south-east Canada.

International dialogue between Canada and the US continued, with a new sense of urgency (at least on the part of the Canadians). Canadian officials awaited the March meeting between Mulroney and Reagan with eagerness; significant progress in cleaning up acid rain seemed to them a real prospect (*NYT*, 15 March 1985, p. 2). US political commentators saw things differently (*NYT*, 16 March 1985, p. 22). In their view, although Reagan *ought* to regard Prime Minister Mulroney as a political ally, he was highly likely to deny him any promise of action on acid rain. The commentators were proved right, and the much-awaited meeting failed to produce any real agreement to control acid rain (*NYT*, 20 March 1985, p. 12). Fears were expressed that the President was bringing US relations with Canada to their lowest point in decades by not agreeing specific control goals or persuading the EPA to tighten and enforce its own regulations (*NYT*, 8 April 1985, p. 17).

The Mulroney–Reagan meeting did have *one* tangible outcome, however. The premiers announced the appointment of a joint team to examine acid rain issues, with Drew Lewis (former US Transportation Secretary) and William Davis (former Ontario Premier) serving as 'special envoys' to the host governments (*NYT*, 18 March 1985, p. 1). The Drew Lewis appointment was, in retrospect, to be a major concession by the Reagan administration, because he emerged as a man of reason and not without stature. It was to be he who eventually persuaded President Reagan that acid rain *was* serious, controls *were* necessary, and further delay *could not* be excused.

Through most of 1985 the President stubbornly refused to accept that there was sufficient scientific understanding of acid rain to warrant action to curb it. This stance was *not* particularly pleasing to the American public. An editorial in the *New York Times* on 30 August 1985 (p. 24) criticized the Reagan administration for having double standards, in accusing the Soviet Union of using chemical warfare resulting in yellow rain in Indo-China (even though scientists disagreed about its cause) and refusing to do anything to prevent acid rain (even though there was plenty of scientific evidence pointing to its cause). Lee Thomas, then EPA Administrator, later pointed out that the Reagan administration *did* acknowledge that acid rain was a serious problem, but it maintained that, while major uncertainties remained, research should continue (*NYT*, 2 October 1985, p. 26).

The 'need for more research' argument was being seen increasingly as a 'stalling for time' tactic, and the patience of many environmentalists and politicians was by now being tested to the full. In October, New York state Attorney-General Robert Abrams somewhat cynically praised the President for his 'masterly irresolution' on acid rain (*NYT*, 27 October 1985, p. 23); two months later, Michael Oppenheimer of the EDF again criticized the President for failing to order reductions in smokestack emissions of SO_2 and NO_x (*NYT*, 23 December 1985, p. 17).

Behind the scenes, Drew Lewis was surveying the acid rain political and scientific battleground, now littered with the remains of outdated arguments and threatened with a serious re-advance of public interest. In September he was to tell governors of six New England states that he believed the acid rain problem *was* caused by sulphur emissions from industry, and that he was about to propose a major clean-up programme (*NYT*, 14 September 1985, p. 1). If action was *not* taken to reduce acid rain *now*, he argued, the ultimate cost of not doing so would far outweigh the estimated $6 billion required to limit SO_2 emissions from factories and power stations (*NYT*, 17 September 1985, p. 30; 22 September 1985, p. 5). Good news indeed for the swollen ranks of opponents to the administration's overt indifference to the problem over the past five years. Lewis's views were eagerly seized upon by the press, which hailed them as a major embarrassment to the President, even though the proposed clean-up programme would be modest in scale.

By the close of 1985, therefore, the President's own adviser on acid rain was in favour of *action* rather than words or further research. The scene was set for significant strides forward in the US acid rain debate through 1986.

MINOR CONCESSIONS (1986)

The early promise of shifting attitudes came to fruition in early January, when William Davis and Drew Lewis – the Canadian and US special envoys on acid rain – submitted their joint report to both governments. The report concluded that acid rain *was* causing serious economic and social problems in the United States, and that acid rain from the US was causing damage in Canada (*NYT*, 8 January 1986, p. 6). It also concluded that acid rain need not be 'researched to death' and that immediate action was required to curb it (*NYT*, 9 January 1986, p. 6). Amongst its main proposals was the speeding up of a five-year $5 billion initiative to develop clean ways to burn coal, with costs shared by federal government and industry. Many observers were concerned but not surprised that the report proposed *no* specific targets for reducing acid rain at source.

Although many Canadian environmentalists expressed disappointment with the report, William Davis stressed that it was the best he could achieve then with the Reagan administration (*NYT*, 9 January 1984, p. 6), which had in any case shifted its thinking on the issue very significantly while the report was being compiled. Canadian Prime Minister Brian Mulroney endorsed the report (*NYT*, 20 March 1986, p. 1), and President Reagan was widely encouraged to accept the proposed solution for the acid rain problem (*NYT*, 12 January 1986, p. 26).

Reagan, ever conscious of the growing public demand for US action and the electoral implications of this groundswell, did endorse the report when he met Brian Mulroney in Washington on 18 March (*NYT*, 13 March 1986, p. 1). Mulroney placed the problem of acid rain at the top of the agenda for the meeting (*NYT*, 16 March 1986, p. 5), and some political commentators believed that, *if* he took a firm enough stand, the Reagan administration would concede to launching an effective programme to reduce industrial and vehicle emissions of SO_2 and NO_x (*NYT*, 18 March 1986, p. 27). The President *did* formally acknowledge at the meeting that acid rain was a serious environmental problem, and that it crossed the border between the US and Canada (*NYT*, 20 March 1986, p. 1). He also accepted the need for immediate efforts to reduce it at source. The President's concessions – even though they were no way near the scale of those offered by most other countries (especially those that had joined the '30 per cent club' – see chapter 8) – were widely welcomed, and his new-found commitment (however reluctantly fostered) was seen as a much-needed help in breaking a serious impasse on acid rain control which had built up in Congress (*NYT*, 23 March 1986, p. 22).

There may have been more than a little self-interest in the Reagan administration's apparent about-turn in its attitude to acid rain control. By early 1986 acid rain had become a serious diplomatic issue between the United States and Canada, and this was starting to affect Reagan's popularity at home and his credibility in the international political arena. But also by early 1986 the US was being threatened with a dose of its own medicine. A large new copper smelter, which had recently been constructed at Nacozari over the border in Mexico, was about to start emitting vast quantities of SO_2, which would inevitably drift across the US–Mexican border and start to pollute parts of the Southwest states (*NYT*, 13 February 1986, p. 30). The Reagan administration could hardly complain about trans-frontier air pollution northwards from Mexico and at the same time turn a blind eye to its own exports of air pollution northwards to Canada.

CONCLUSION

Recent years have seen mounting concern about the spread of acid rain within the United States. What started in the early 1970s as a problem initially confined to the Northeastern USA (and to the Adirondack Mountains in particular), had by the early 1980s become a national problem. Eastern states blamed states in the West and Midwest for exporting their air pollution downwind, and a long-running political debate with all the acrimony and intrigue of a soap opera got under way.

The US also appeared to be exporting vast quantities of SO_2 (the basic ingredient of acid rain) across its northern borders, into south-eastern Canada. This sparked off an equally long-lasting political and diplomatic series of exchanges between the USA and Canada. Canadian frustration that the USA was not taking the issue seriously rose steadily, and, despite mounting scientific evidence (from within the USA and Canada, as well as from European research) that SO_2 emissions from power stations and industry were a significant contributor to the production of acid rain, the Reagan administration has until recently persisted in the belief that more study and research were required.

The recent about-turn in government attitude – reflected in Reagan's acceptance (however reluctantly) of the Lewis/Davis report – *does* mark a significant watershed in the official position. After years of mounting scientific evidence, campaigning from environmental groups, and rising general public awareness of and concern about acid rain in North America, the first tentative steps have been made towards a satisfactory solution. It remains to be seen whether the administration's concessions turn out to be *too small* and *too late*. Certainly, when viewed alongside the commitments to major reductions in emissions of SO_2 and NO_x that have been made by many countries over the past five years, the US proposals seem conservative in the extreme!

10 ACID RAIN IN BRITAIN
'Not guilty, m'Lord . . .

> Acid rain is thus a real problem, but its importance has often been exaggerated. It does much less harm in the populous parts of Britain than did the air pollution we formerly suffered. It has, however, been taken up as a *cause célèbre* by the so-called 'ecologists' who were disappointed because their grim prophecies of doom from pollution and starvation have not been fulfilled.   (Kenneth Mellanby 1984b)

Acid rain is 'now the most controversial form of air pollution in the developed world' (Myers 1985: 113). Britain has *not* escaped the controversy, and views on acid rain differ markedly, even amongst the experts. There are those like Kenneth Mellanby who regard the issue as a diversion of attention away from more pressing environmental issues. True, there are presagers of ecological doom who readily seize upon the 'conveniently evocative catchphrase' of acid rain (*The Times*, 20 August 1982, p. 2). But there are also those who share the Royal Commission on Environmental Pollution's view of acid rain as 'one of the most important pollution issues of the present time' (1984). Unfortunately, scientific opinion by and large sides with the Royal Commission, and government opinion – in Britain at least – sides with Mellanby.

In this chapter we shall explore how these two groups have interacted since 1982 to produce a political debate of national and international significance, as contentious as it has been prolonged. The debate has seen scientist set against scientist, scientist set against politician, and politician set against politician. Political moghuls have faced a barrage of opposition from a curious and highly unusual alliance between some government scientists and the environmental lobby in Britain, and from leaders of most European countries, all of whom have grown increasingly frustrated by the Thatcher government's overt reluctance to grasp the nettle of reducing sulphur dioxide emissions.

THE EARLY DEBATE

Mid-1982 represents a major watershed in the attitude of the British government to acid rain. Up to that time the official view was 'there is no case to be answered'. Department of the Environment (DoE) officials and Whitehall mandarins basked in the glory of the visible successes of anti-smoke legislation introduced in the 1956 Clean Air Act (described in chapter 1). Declining smoke concentrations in cities like London (Figure 10.1), which admittedly had been declining for some time *before* the 1956 Act, brought cleaner buildings, brighter skies, and a healthier population. By mid-1982, however, there were signals – albeit weak ones – of a growing government concern over mounting international pressure to reduce oxide emissions (chapter 8) and accumulating scientific evidence of the causes and impacts of acid rain in the environment (chapters 4–6).

The prime catalyst for this new-found concern was the international meeting of scientists held in Stockholm in June 1982 (*The Times*, 30 June 1982, p. 3) – ten years after the UN Conference on the Human Environment – to evaluate the causes and effects of acid rain across the world (chapter 8). Britain and the United States were singled out for serious criticism as major sources of contaminants, and the preliminary results of the OECD surveys on trans-frontier air pollution carried out in the mid-1970s (Figure 2.3; chapter 8) were shown to be substantially correct. Between the mid-1950s and the mid-1970s atmospheric concentrations of sulphur across Europe as a whole had risen on average by 50 per cent. In Scandinavia and Central Europe they had doubled. Britain stood accused of exporting air pollution overseas, and was drawn reluctantly into the international debate on acid rain. In fact Britain was projected to a high-profile position in centre stage, where it has remained (with the United States, chapter 9) ever since.

The British government's attitude to acid rain has – like that of the United States (chapter 9) – been nothing if not cautious. Amongst the government's limited repertoire of defences has been the call for more research before taking binding decisions – a tactic used recurrently since 1982. Mr Patrick Jenkin (Secretary of State for the Environment) underlined the importance of research in a speech to the House of

Figure 10.1   Declining average winter smoke levels in London, 1921–71

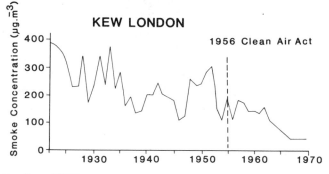

Source: after Rees (1977)

Commons on 7 March 1984: 'it really is necessary to establish a clear idea of the cause and effect before spending millions of pounds which might turn out to be useless' (*The Times*, 8 March 1984, p. 4). However, the strategic benefits of creating breathing space, particularly when accusations are coming in from all quarters, have not been overlooked by critics who argue that 'conservation measures sometimes have to be taken on suspicion, for fear that the patient may die before a conclusive diagnosis is made' (*The Times*, 10 January 1984, p. 11).

The 'need more research' policy is evident, for example, in the reversal in June 1982 of an earlier government decision to cut research funding on acid rain (*The Times*, 30 June 1982, p. 3). The DoE was told by the Cabinet to reinstate a £120,000 grant to the Institute of Terrestrial Ecology (ITE) at Monks Wood, near Cambridge, to carry out research on the extent of damage to crops, forests, and lakes in Britain from acid rain. This timely and highly visible about-turn in overt government commitment was doubtless *not* unrelated to the impending publication of a survey of the extent of acid rain in Britain commissioned by the DoE.

The first substantive evidence of acid rain falling on home ground in Britain appeared in late August 1982. It came in the preliminary report of the UK Review Group on Acid Rain, based at the government's Warren Spring Laboratory in Hertfordshire. The report concluded that the acidity of rain in some regions appeared to be 'of the same order as in the high input areas of Scandinavia and North America' (1982). The survey, carried out between 1978 and 1980, covered Scotland, the North of England, the East Midlands, and parts of southern England and Wales, and it found that dry deposition of sulphur 'is comparable with or exceeds wet deposition in large areas'. It recommended more measurements in southern England, most of Wales, and Northern Ireland to complete the picture of spatial variations, and more effort 'to assess the importance of rain acidity'.

The UK Review Group report was to be the first of many demonstrating that Britain *does* have a significant role to play in the acid rain debate, implicated as a producer of the sulphur and nitrogen oxides believed to be associated with the production of acid rain that falls in Britain and downwind overseas (especially Scandinavia).

International initiatives

Acid rain was starting to gain its political colours in Britain when the Geneva Convention on Long-Range Transport of Air Pollutants (chapter 8) came into force on 15 March 1983, with little ceremony (*The Times*, 17 March 1983, p. 6). Both the United States and Britain were reluctant signatories of the treaty (Table 8.5). But the attitude of the British government was *not* seriously challenged by the Convention, which is simply an accord for collaboration in research and for informing each country of new industrial projects that might add to the sulphur dioxide budget. It did *not* set emission standards or impose sanctions against

recurrent offenders. Neither did it take any direct action against countries
– like Britain (Figure 10.2) – that were clearly exporting air pollution to
their downwind neighbours.

Although the Prime Minister was apparently 'let off the hook' this time,
leading questions about hidden agendas in the debate were being asked
and public opposition to the government's overtly recalcitrant attitude to
acid rain was mounting. The gauntlet was thrown at the feet of both the
CEGB and the British government by critics who noted that countries like
Britain and the United States manage to disperse pollutants to the
detriment of others without themselves suffering discharges from elsewhere. As a result, it was argued, 'they are disinclined to pay for the
modifications that would cut gaseous emissions to levels which the victims
of transboundary pollution, notably Scandinavia and Canada, would
like. . . . Technical means are available, and the net cost is not inordinate.
Britain should be ready to play its part in internationally agreed measures'
(*The Times*, 15 March 1983, p. 13).

The CEGB, cast as the villain of the piece, was first drawn into the

Figure 10.2   Trajectories of air masses over Western Europe

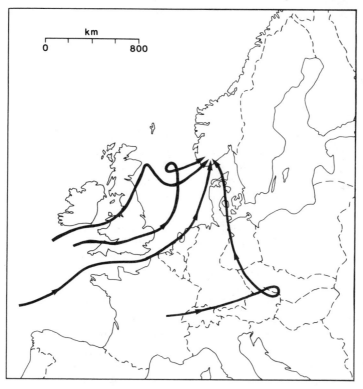

Source: after Lunde (1976)
Note: Based on plume tracking of four air masses, by airplane, in September 1973

public debate by its Secretary John Anderson, who shouted 'foul play' and dismissed as unfair an editorial in *The Times* on 2 April 1983 that 'paints a picture of considerable reluctance on the part of the British government and industry to comply with the proposals of the UN Convention. The opposite is in fact the case' (*The Times*, 2 April 1983, p. 7). The CEGB put forward various lines of defence (reviewed on pages 238–40), which were quickly to become enshrined in the 'institutional' view of acid rain. John Anderson insisted that there was *no* foundation to the claims of lack of interest in acid rain by the CEGB, which 'has always been anxious to ensure that the environmental effects of its operations give no cause for concern. Its scientists are currently assisting Government departments to ensure that the United Kingdom plays a full part in implementing the UN Convention. However, it is essential that this co-ordinated effort is directed towards measures which provide effective control at a price which is commensurate with the environmental benefits' (*The Times*, 2 April 1983, p. 7).

By Easter 1983 the cat and mouse game had started, and the sides were lining up for a prolonged exchange of acid rhetoric. Within a year the pace of the debate had started to speed up, albeit slowly and erratically. Government diehards, ever eager to quieten if not discredit the environmental lobby, continued to insist that 'the politics of acid rain have run ahead of the science' (*The Times*, 7 May 1983, p. 8).

Research requirements

Uncertainty about the detailed aspects of causes and effects was by now rising as the first wave of detailed scientific research was discovering just how little the problem was then understood. For example, experts at a Royal Society of Chemistry Symposium on Acid Rain, held in April 1983, were sharply divided on what might be done about acid rain because of the complexity of the chemical reactions within the atmosphere that led to its formation (*The Times*, 13 April 1983, p. 2). This fortunate revelation of ignorance by the enlightened gave the government – ever happy to seize upon such timely public displays of lack of consensus amongst scientists with sound credentials – a welcomed opportunity to buy breathing space by investing in further research. In May, the DoE announced that it was allocating nearly £500,000 during that financial year for research into acid rain (*The Times*, 7 May 1983, p. 8).

The time-honoured tactic of gaining time by funding more research was used most visibly on 5 September 1983 when the Royal Society announced a £5 million study of 'the causes of acidification of surface waters in affected areas of Norway and Sweden' (*The Times*, 3 September 1983, p. 6; 6 September 1983, p. 2). The five-year project was to be managed jointly by the Royal Society and the Academy of Sciences in Norway and Sweden, under the leadership of Sir John Mason, Director General of Britain's Meteorological Office. Funding – described by some critics of official British attitudes to acid rain, at home and abroad, as contaminated conscience money – was provided by the CEGB and the National Coal Board (NCB).

The sponsors proposed that the study should aim to answer four fundamental questions:

(1) What factors, in addition to acidity, affect fisheries in the lakes of Norway and Sweden?
(2) What improvements in the chemistry of surface water would come from reductions in man-made sulphur emissions?
(3) What levels of acidity can various fish species tolerate?
(4) How do the biological, chemical, and hydrogeological characteristics of catchments influence the composition of water quality?

Within this relatively massive research budget, around £600,000 was to be allocated to a project based at Loch Fleet in Galloway (south-west Scotland) designed to re-establish fisheries in affected waters and to explore techniques of land treatment and management to improve water quality and reduce acidity (*The Times*, 21 March 1984, p. 4). There was mounting evidence of acidification of freshwater lakes and streams in this part of Scotland (chapter 4), and Loch Fleet provided a good opportunity for evaluating possible solutions.

The vested interests of the sponsors, eager to associate themselves with attempts to resolve acid rain problems but even more eager to dissociate themselves from any implied responsibility in creating acid rain, is writ indelibly on the four questions. It is highly significant that the study was *not* to be concerned with the two key areas on which the British government had sought detailed information and on which the scientific community could as yet offer only partial answers (chapter 3) – the complex processes that create acid rain, and the degree to which reductions in emissions of SO_2 from coal-fired power stations in Britain might alter the acidity of rainfall over parts of Norway and Sweden. The study was quickly denounced by Norwegian scientists as an exercise in rediscovering the wheel (*The Times*, 8 June 1984, p. 16).

As an exercise in international diplomacy, the Royal Society research study was designed partly to convince Britain's Nordic neighbours that it was taking seriously the accusations that emissions of oxides from British power stations played a significant role in creating acid rain in Scandinavia. However, in January 1985 Swedish scientists refused to become pawns in the international political debate by rejecting a 1 million Swedish kroner (about £100,000) grant to study the effects of acid rain in Sweden, and engaging the attention of the media in doing so. 'The main purpose of the British offer', said Mr Sten Bergstrom of the Swedish Meteorological Institute, 'is to buy time so that Britain can continue to spew smoke all over Europe.' He added that it would be morally wrong to accept funds offered by the British power industry in view of Britain's refusal (in December 1984) to join the '30 per cent club' and reduce sulphur emissions (chapter 8); 'we can't accept that, since it is our duty as scientists to protect the Swedish environment' (*The Times*, 19 January 1985, p. 5).

Domestic damage

As acid rain overtures were being played across Europe (chapter 8), the British government continued to play Nero's fiddle whilst West German forests died (chapter 5). But problems of ecological change attributed to acid rain were soon to be mentioned in dispatches from home. Many were included in a report on the effects of acid rain on plants and animals in Britain, produced by the Nature Conservancy Council (NCC) and described as 'alarming' (*The Times*, 8 June 1984, p. 16). The report was drafted in May 1983 (*The Times*, 1 November 1983, p. 16) but remained unpublished until mid-June 1984 (*The Times*, 23 June 1984, p. 9), after the House of Commons Select Committee on the Environment began what was to become a critical inquiry into acid rain in Britain. It is highly unlikely that the draft NCC report was neither studied nor delayed in Whitehall – its startling findings doubtless explain both the long delay in and critical timing of publication for public consumption.

The NCC survey (Fry and Cooke 1984) found that rainfall readings below pH 4.6 (regarded as very acid) were common throughout Britain, and in eastern areas they averaged less than 4.3 (even more acidic). It compiled a long and sombre catalogue (Table 10.1) of ecological changes and damage believed to be associated with acid rain. For example, some attempts by the Forestry Commission to establish new coniferous

Table 10.1   *Some effects of acid rain on wildlife in the UK*

Wildlife affected		Area
Dippers (riverside birds)	–	decline in south-west Scotland, north-east England, Wales
Char (fish)	–	south-west Scotland, threatened with extinction in Loch Doon
Otters	–	declining in many rivers in Wales (e.g. Tywi, Irfon, and Tyfi) as fish (esp. trout) disappear
Natterjack toad	–	vanished from Surrey ponds
Lichens	–	declining in many areas
Mosses/peat bogs	–	decaying in mid-Wales (south of River Wye)
Brown trout	–	declining in south-west Scotland
Common frog	–	altered distribution in south-west Scotland
Smooth newts	–	altered distribution in New Forest
Freshwater shrimps	–	vanished from Ennerdale Water in the English Lake District
Freshwater limpet	–	probably vanished from River Duddon in the English Lake District

Source: Fry and Cooke (1984)

plantations in the Pennines had failed because of the effects of sulphur pollution. The survey reported that 'phytoplankton invertebrates, bog vegetation and fish seem to have been affected in a similar way to that in Scandinavia and North America – a loss of species diversity and increased abundance of acid tolerant species has been recorded' (*The Times*, 1 November 1983, p. 16). The report also pointed out that, of 357 sites in the UK listed as of international importance for nature conservation by the NCC in 1977, over 100 were in areas with geology especially vulnerable to acidification. Many of them were likely to be seriously damaged if acid rain was not reduced.

Professor F. T. Last of the Institute of Terrestrial Ecology at Edinburgh also reported a number of areas in Scotland where fish populations had been seriously reduced through acid rain – including Galloway, upper stretches of the River Tay, and the headwaters of the Forth. The most vulnerable areas were 'catchments afforested with evergreen conifers growing in acid soils, overlying slowly weathering bedrock in areas receiving large amounts of acidic deposition' (*The Times*, 1 November, 1983, p. 16). Similar reports of declining fish populations were starting to arrive from north and west Wales, particularly from areas draining coniferous forests (see chapter 4). By November 1983 the Welsh Water Authority had embarked on the routine monitoring of rainfall acidity (*The Times*, 1 November 1983, p. 1).

Whilst the draft NCC report was collecting dust on the shelves of Whitehall, the finishing touches were being put to a detailed report on acid deposition in the United Kingdom prepared for the DoE by Warren Spring Laboratory, which appeared on 9 January 1984. It had two main conclusions: there *was* evidence of an increase in rainfall acidity in the UK, but the lack of research data precluded any detailed identification of its sources (Barratt and Irwin 1983). The report suggested that emissions of SO_2, mainly from coal- and oil-fired power stations, factories, and refineries, were becoming relatively less of a problem than increases in nitrogen oxide concentrations (from vehicle exhausts) (Figure 10.3) – both had increased considerably this century, but SO_2 concentrations peaked in the mid-1960s and had declined since 1970. The government responded to the report on 9 January with a commitment to fund further long-term research – described by Friends of the Earth (FoE) as 'utterly inadequate' (*The Times*, 10 January 1984, p. 2).

The Warren Spring report was *not* well received outside government circles. It embodied the scientist's traditional concern for detail, inherent conservatism, and extreme reluctance to read too much into the available evidence. Critics claimed it did little if anything to clarify the suggested link between sulphur emissions and the death of trees and fish; it dwelt on inadequacies of earlier data and on the large number of factors to be taken into account; and that it ended with the now customary call for further research 'over an extended period (decades)' (*The Times*, 10 January 1984, p. 11).

Figure 10.3   Nitrogen oxide emissions in the UK, by source, 1972–82

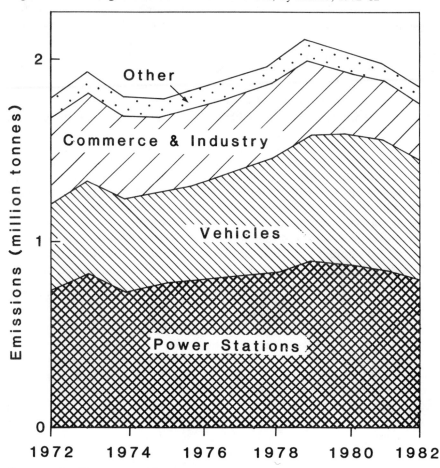

Source: after Department of Environment (1983)

Parliamentary debate

A month after the appearance of the Warren Spring report, acid rain was being discussed in parliament (*The Times*, 9 February 1984, p. 4). When asked by Mr Robin Corbett (Labour member for Birmingham, Erdington) to publish a document containing the latest evidence on acid rain in Britain, the Secretary of State for the Environment (Mr Patrick Jenkin) replied that the amount of money being spent on research into the effects of acid rain was being substantially increased and that a briefing paper had been prepared for the Select Committee on the Environment, which would shortly be investigating acid rain. Challenged to 'stop being so complacent' and asked would he 'press for further international cooperation to avert this menacing environmental matter', Mr Jenkin vigorously refuted any charge of complacency and charged his Labour interrogator of 'making a large number of assumptions which scientific evidence does *not* support'.

Within a month acid rain was once again to appear on the agenda of the Palace of Westminster, in the wake of the publication on 23 February of the 10th Report of the Royal Commission on Environmental Pollution, which classed acid rain as 'one of the most important pollution issues of the present time' which was at the forefront of environmental debate in Europe (*The Times*, 23 February 1984, p. 13). In the Commons on 7 March the Secretary of State was asked by Mr Malcolm Bruce (Labour member for Gordon) if he would give serious consideration to the early acceptance of the Royal Commission's recommendations on acid rain 'since it appears that we are being affected by deposition' (*The Times*, 8 March 1984, p. 4). Mr Jenkin again fell back on the need for further research, noting that £1 million had recently been added to the research budget for acid rain.

However, for the first time in public debate the private fears of the environmental lobby were anticipated when the spectre of a nuclear alternative to conventional power stations was raised by Mr Jenkin, who noted that 'in the longer term, the Royal Commission recognizes that one way to deal with and achieve a reduction in emissions of sulphur dioxide is by a modest increase in nuclear power, and I think this is something the Government would welcome' (*The Times*, 8 March 1984, p. 4).

Friends of the Earth and other environmental pressure groups recognized the need to tread very carefully in presenting their case for the installation of equipment that would reduce SO_2 emissions from power stations and at the same time increase production costs for electricity. They faced the dilemma 'that cutting acidic discharges might decisively alter the economics of power supply in the nuclear option's favour. Having to make choices between an inferred source of pollution and a potential one is unsatisfactory' (*The Times*, 10 January 1984, p. 2). It was a welcome sign that MPs in the all-party House of Commons Select Committee report on acid rain showed *no* eagerness to see sulphurous coal replaced by more nuclear power (*The Times*, 7 September 1984, p. 5).

THE SUMMER OF '84

The miners' strike during most of 1984 highlighted the nation's traditional and continued reliance on coal as a source of energy. Although the CEGB's output of electricity was *not* seriously affected by the coal shortages (partly because the Board could switch the main production to oil-burning power stations), the government's resolve to reduce the nation's dependence on coal (and, more specifically and ideologically relevant, its prospects of being 'held to ransom' by militant trade union activity) was strengthened by the bitter, costly, and protracted experience.

The miners' strike also had a more immediate impact on government attitudes to acid rain, because it effectively monopolized the energies and attention of Cabinet ministers and Whitehall officials over the period when the political debate over acid rain was raging most fiercely throughout mainland Europe (chapter 8). George Munroe, press spokesman for the Department of Energy, told journalists early in October that neither of the department's two new ministers (Under-Secretary David Hunt and Parliamentary Under-Secretary Alistair Goodland) was available for interviews on acid rain because 'all their time is taken up with the Coal Board and energy conservation' (*The Times*, 4 October 1984, p. 14).

The summer of 1984 was to see three particularly important developments in the acid rain saga in Britain: the prospect of a marked switch in government attitude to the issue, the preparation and submission of a House of Commons Select Committee report on acid rain, and the realization by the scientific community of the complexities of the problem.

Prospects of change

The first signs of a possible change in government attitude came over the Spring Bank Holiday towards the end of May, when an unprecedented series of briefings on environmental issues took place during the Prime Minister's weekend retreat at Chequers. The session was called after Helmut Kohl, West Germany's Chancellor, had complained to the British government about its reluctance to join its EEC partners in reducing acid rain (chapter 8).

A group of distinguished scientists including Sir Herman Bondi (Chairman of the Natural Environment Research Council), Dr Martin Holdgate (Government Chief Scientist at the DoE), Sir John Mason (Director of the Meteorological Office), and Dr Peter Chester (CEGB Director of Research) spelled out to government scientists and ministers – including Mrs Thatcher – the effects of acid deposition on aquatic ecosystems and forests, the atmospheric processes involved, and the various control technologies available (*The Times*, 8 June 1984, p. 16). The DoE was in favour of remedial action, but Walter Marshall of the CEGB convinced the Prime Minister that there was *no* reliable evidence that cutting emissions of SO_2 from power stations in Britain would save trees in West Germany, fish in Scandinavia, or cathedrals in Britain.

Political commentators quickly formed the belief that, given the mounting international pressure on Britain and the accumulating evidence of acid rain impacts in Britain (and in particular in parts of Scotland), the Prime Minister was at long last preparing to put right the neglect that British environmental policy had suffered over the previous five years. Hopes were high that she would announce a reversal in British policy on acid rain when she attended the European summit scheduled to take place in Paris in June, and announce Britain's accession to the so-called '30 per cent club' (described in chapter 8). Tony Samstag of *The Times* underlined the momentous nature of the expected change of heart: 'it is all a bit as if

Mr Heseltine had decided to declare the Greenham Common air base a nuclear free zone' (8 June 1984, p. 16).

However, the speculations of a thaw in political opinion were *not* to be reflected in the government's public activities in the weeks and months ahead. In a parliamentary debate on the EEC Directive on Air Pollution on 8 June, Mr William Waldegrave (Under-Secretary of State for the Environment) stressed that 'the Government shared the view with other EEC countries that it is essential to identify and overcome the causes of the damage to forests and water, but it was difficult to disregard the scientific uncertainties' (*The Times*, 9 June 1984, p. 4).

In the same debate the mood of many scientists and environmentalists, and a growing number of politicians (of all parties), was captured by Mr David Clarke, an opposition spokesman on the environment, who said that the government's approach to the EEC Directive was the all too familiar story of it saying it recognized the problem but did not intend to do anything about it. The government offered its by now stonewall reply that it would be costly to tackle the problems of acid rain (at home and abroad), and the CEGB would have to increase electricity prices by 10 per cent to meet the costs involved. Mr Nigel Spearing (Labour member for Newham South) replied somewhat caustically that 'when you get down to it the Chicago School of Economics and the proper concern for the protection of the environment do not mix' (*The Times*, 9 June 1984, p. 4).

Government Select Committees

Whilst their parliamentary colleagues were listening to the debate in Westminster, the all-party House of Commons Select Committee on the Environment had travelled *en bloc* for a five-day visit to study environmental damage attributed to acid rain in Norway and Sweden. Over the next three months the Select Committee was to hear evidence submitted by all the main groups with an interest in acid rain in Britain, including government departments and quangos, local authorities, industry, scientists, water authorities, environmental groups, trade unions, doctors, farmers, architects, and surveyors. As the Commons Committee embarked on its inquiry, the House of Lords Select Committee on the European Communities finished its study of acid rain in Britain. The Lords came out against the government's position on the issue and in favour of Britain joining the '30 per cent club' (House of Lords Select Committee on the European Communities 1984).

The Commons Committee was bombarded with facts and figures, theories, and speculations. Many cases were impassioned if not dogmatic – the stakes were high. Friends of the Earth accused the CEGB of indifference to the issue; the CEGB argued that it was being wrongly accused and that high concentrations of ozone (produced by the emission of nitrogen oxides from vehicle exhausts) were now widely held to be a significant cause of Europe's dying forests (*The Times*, 12 June 1984, p. 2).

The Committee was told by the DoE that emissions of SO_2 in Europe had doubled since 1950, but UK emissions had fallen by 12 per cent over that time; the UK contribution to European air pollution had dropped from 25 per cent to 11 per cent over those three decades (*The Times*, 12 June 1984, p. 2). Scientists from the Freshwater Biological Association laboratories at Windermere told the Committee that no increase in acidity in rainfall, rivers, or lakes had been recorded in the Lake District in the past thirty years, and that there was no direct evidence to link observed ecological changes in trees and lichens with acid rain (*The Times*, 5 July 1984, p. 3).

Many cases presented to the Committee, both for and against emission control, stressed the lack of direct evidence to link possible causes with observed patterns of acid rain, to link acid rain with observed changes in forests, lakes, rivers, and soils, and to link proposed reductions in SO_2 emissions with anticipated reductions in rainfall acidity and impacts on sensitive areas.

Scientific evidence

It is probably *no* fortuitous coincidence of timing that between June and August 1984 – whilst the Select Committee was sitting – a number of important scientific reports germane to acid rain appeared in Britain.

Late June saw the publication of a report of a fifteen-month study on acid rain commissioned by the Department of Energy and carried out by the Energy Technology Support Unit at Harwell, Oxfordshire. This argued that the degree of blame attributed to British power stations and factories was misplaced (Buckley-Golder 1984). It showed that Britain produced much less acid rain that drifted eastwards towards Europe than had often been alleged – their figures suggested that less than 2 per cent of the acid rain falling on Norway and Sweden came from Britain, and frequent anticyclones over central Europe produced circulations of air that carried about half of Europe's oxides over Scandinavia. The report also ascribed much of the forest decline in West Germany and elsewhere to ozone. Whilst it conceded that the mechanisms of injury to trees were *not* proven, the belief was that ozone absorbed by foliage damaged the cuticles of leaves, so that nutrients were leached from the tree when it rained and the tree quite literally starved to death.

The British delegation to a meeting of EEC environment ministers in Munich in late June 1984 (chapter 8) was fully armed with this Department of Energy report and fully prepared to argue the case against forcing Britain to make binding commitments to reduce SO_2 and NO_x emissions within a fixed period, particularly whilst there was still so much speculation surrounding acid rain (*The Times*, 23 June 1984, p. 9). Unfortunately for the British government and its delegation, the NCC report of acid rain – which had been delayed for over a year, and was much more critical of the CEGB than was the Harwell report – was published during the week of the Munich meeting.

SELECT COMMITTEE REPORT

The much-awaited report of the House of Commons Environment Committee appeared in September 1984, with conclusions and recommendations that were 'likely to embarrass the Government and CEGB' (*The Times*, 13 August 1984, p. 2). The report 'received a predictably warm welcome from environmental groups and from the Labour, Liberal and Social Democrat Parties', although 'there was surprise in some quarters that the committee had come to such forthright and categoric conclusions' (*The Times*, 7 September 1984, p. 5).

The Committee pointed the finger squarely at the CEGB in concluding that emissions of SO_2 from British power stations were one of the main causes of acidity in thousands of European lakes, including many in Galloway (south-west Scotland). It did, however, acknowledge that emissions of NO_x from vehicle exhausts, ozone, and chemical reactions in the atmosphere could significantly contribute to environmental damage. The tone of the report was well captured by Tony Samstag of *The Times*: the 'committee declared itself "deeply disturbed" by Britain's policy on acid rain, and "appalled" by failure to monitor the damage done to buildings by corrosion from air pollution. The absence of serious research into the phenomenon had been "a major failure", and evidence given by the CEGB, arguing against many of the remedial proposals, "trite and evasive" ' (3 December 1984, p. 3).

The Committee's members (representing all the major political parties) were unanimous in recommending that Britain should join the so-called '30 per cent club' (chapter 8) without delay. They also recommended that Britain should back the draft EEC Directive reducing sulphur emissions by 60 per cent by 1995, even though this would require the CEGB to install flue gas desulphurization (FGD) equipment (chapter 7) at ten coal-fired power stations (Figure 10.4) by 1995. This would cost an estimated £1,500 million to install and £350 million a year to run (and force electricity price rises of up to 5 per cent). The Select Committee also expressed its belief that the best way to reduce pollution from cars was to develop 'lean burn' engines (rather than use special catalysts) of the sort presently being manufactured in Japan, which produce much less pollution (chapter 7).

Hopes were raised that, with the Select Committee's guidance and thrust, the government might now grasp the nettle and act on acid rain. Inevitably the Committee's proposals were not welcomed by the government, which four months earlier had refused to join the '30 per cent club' and whose junior environment minister (Mr William Waldegrave) had told the Committee that it believed the case against SO_2 emissions was not yet proven and the government was not prepared to spend hundreds of millions of pounds in reducing SO_2 pollution until there was greater certainty about the resultant benefits. This same view persists largely unmodified today, and it mirrors official US attitudes under the Reagan administration (chapter 9).

Figure 10.4   CEGB power stations selected for fitting of FGD equipment to meet the EEC Directive for a 60 per cent cut in SO$_2$ emissions between 1980 and 1995

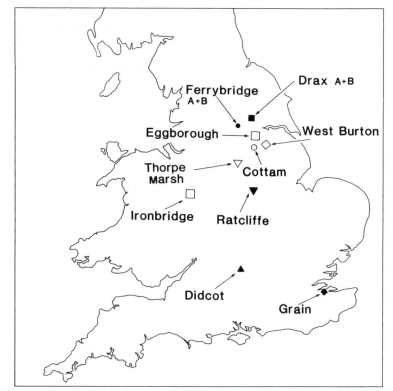

Source: after Anon (1984b)

The report had its critics outside as well as inside government. Many had reservations about the report's 'undiscriminating approach', which made 'little attempt to apportion blame for different effects, or to match either the forms or costs of its proposals with the evils to be remedied' (*The Times*, 7 September 1984, p. 5).

Publication of the Select Committee report gave a much-awaited opportunity for interested commentators to air their views on government attitudes towards acid rain. Friends of the Earth were quick to point out that electricity prices in Britain had in fact risen by 45.5 per cent between 1973 and 1984, so that a further rise of 'ten per cent over ten years appears a small price to help save our cathedrals, lakes, rivers and forests' (*The Times*, 11 September 1984, p. 15). They also indicated that a cheaper and more effective solution would be to trim 5 per cent from the CEGB's projected rise in electricity consumption by energy conservation schemes (chapter 7). This, they argued, would allow Britain to implement the

EEC's proposed 60 per cent reduction in SO_2 emissions at a cost of 3–4 per cent over ten years.

Mr Allan Roberts, Labour MP for Bootle and a member of the Select Committee, underlined the need for urgent action by government. He emphasized the fact 'that acid rain contributes significantly to the environmental damage we witnessed in Europe and Britain is *not* proven, in the same way that it is *not* proven that smoking causes lung cancer. If we wait for the kind of proof the CEGB wants, it could well be too late. The smoking patient may be dead' (*The Times*, 12 September 1984, p. 11).

Criticism came not only from the opposition parties. Mr Michael Lingens, Chairman of the Tory Bow Group, was also extremely critical of the government's cautious approach to acid rain, arguing that it appeared to underestimate the political importance of taking action now, especially in light of the public's willingness to pay for environmental protection via electricity price rises. He concluded that 'the Government's position of tamely trailing along at the back of the international pack is unlikely to prove acceptable for much longer with the public, or to an increasingly conservation conscious party' (*The Times*, 11 September 1984, p. 15).

Singled out for special treatment was the CEGB, whose response to the Select Committee's findings and recommendations was described as 'predictably inadequate' (*The Times*, 2 November 1984, p. 13), especially because it attempted 'to discredit the Select Committee Report by selective quotations which distort what is actually said' (*The Times*, 12 September 1984, p. 11). The CEGB case was founded partly on two main arguments. One was that NO_x played a major part (along with hydrocarbons, ozone, and sunlight) in the observed forest damage in many areas (a point the Select Committee itself made). Its other argument was that the development of new designs for new coal-fired power stations using processes such as fluidized bed combustion (chapter 7) was now in hand, in line with the Committee's recommendations.

But Mr Chris Smith, Labour MP for Islington and Finsbury and a member of the Select Committee, argued that 'the CEGB have totally failed to accept that serious damage is being caused to lakes, streams and fishlife – both here in Britain and abroad – by sulphur dioxide emitted from our power stations' (*The Times*, 2 November 1984, p. 13). The CEGB proposals would still leave the problem of existing power stations (which produced more SO_2 than any other *country* in Western Europe). The Select Committee concluded that the fitting of FGD equipment to existing power stations was the only realistic way of achieving major reductions in SO_2 emissions. Mr Smith protested that 'both the CEGB and the Government have to date ducked this issue . . . surely it is about time Sir Walter Marshall [CEGB Chairman] stopped lashing out at the Select Committee and started considering seriously and carefully those parts of its report he has hitherto dismissed or ignored'.

STORM CLOUDS GATHERING

The acid rain clouds had been gathering over Britain for three years, and by late 1984 many forces were lined up against the government's policy of cold contempt, including the Lords and Commons Select Committees on the Environment, the Royal Commission on Environmental Pollution, the Nature Conservancy Council, and most European governments (chapter 8).

Disenchantment was beginning to set in in the Tory party by the time of their annual conference in Brighton in October (*The Times*, 10 October 1984, p. 4). Acceptable noises had been made by Mr William Waldegrave (Under-Secretary of State for the Environment) when he announced that Britain had reduced its emissions of SO_2 by nearly 40 per cent since 1970 (Figure 1.5) and the government intended a further drop of 30 per cent (from 1980 levels) by the late 1990s. But this was *not* enough to placate Tory MPs. Mr Michael Willis (member for Brentwood and Ongar) proposed and the conference carried a motion calling on the government to conserve the environment by all practical means, including the reduction of known sources of pollution. Sir Hugh Rossi (member for Hornsey and Wood Green), Chairman of the Commons Select Committee on the Environment, argued that Britain *must* join other European countries in reducing acid rain damage, because the evidence strongly suggested that Britain remained the second largest producer of SO_2 in Europe after the Soviet Union (Figure 8.1). Within a month of the Brighton Conference the government was to face even stiffer opposition to its stance on acid rain.

The acid rain storm broke on Monday, 3 December 1984, which was to be a momentous day in the history of the Thatcher government's environmental policies. Mr Waldegrave, ever the bringer of bad tidings, announced in the Commons that the government was rejecting the Select Committee's recommendations on acid rain. The government's detailed reply to the criticisms of the Select Committee (HMSO 1984) marked an abrupt and unpopular, though not entirely unexpected, reversal in what had seemed to be a gradual conversion of the Thatcher government towards the European stance on acid rain. Mr Patrick Jenkin (Secretary of State for the Environment) told parliament on 3 December that the proposed reductions in SO_2 emissions from power stations were simply unacceptable. Three days later Mr Waldegrave was to tell a meeting of European environmental ministers in Brussels that the proposed EEC measures would be too costly. The British government subsequently defended its decision not to reduce SO_2 emissions at a meeting of ministers of the industrial nations (Canada, France, Italy, Japan, the UK, United States, and West Germany) at an environmental summit meeting held in London on 17 December.

The government's retrenched position was to prove unpopular with British MPs and MEPs (Members of the European Parliament). A Greenpeace survey of MPs' willingness to help in any of ten specified environmental campaigns showed that more than one-third of MPs were seeking action to combat acid rain (*The Times*, 17 December 1984, p. 2). Alarmingly for the government, proportionately more Tory MPs than Labour MPs showed signs of concern. Political feelings in Europe were also running high in the wake of the announcement of 3 December, and a backlash by Tory MEPs was planned at a private meeting in Brussels on 5 December (*The Times*, 7 December 1984, p. 5). Thirty of the forty-five British members decided not merely to join, but – in a symbolic act of defiance – to head the great majority of MEPs of all parties and all countries who were hostile to the British government's position on acid rain.

This highly visible and politically calculated act – the first time that Tory MEPs had ranged themselves with majority opinion in Europe in outright opposition to British government policy – was in defence of what they believed to be Britain's best interests. They argued that the EEC Directive was overwhelmingly popular in other member countries, and that the government's opposition to it would do Britain grave danger (especially with public and political opinion in West Germany). At the debate in the Strasbourg parliament on 13 December, the uncomfortable alliance of British Tory MEPs and their European colleagues called on member governments to withdraw their more extreme objections to the draft EEC Directive. The beleaguered British representative, William Waldegrave, was under strict instructions to veto the draft Directive as contrary to British national interest.

The government's conclusion that it would not be right or responsible to rush into a crash programme costing around £1,500 million to install emission control equipment in power stations was debated in parliament on 11 January 1985 (*The Times*, 12 January 1985, p. 12). Mr David Clark (Labour MP for South Shields) accused the government of 'waging chemical warfare' on Britain's countryside and neighbours, and of acting in a 'wantonly irresponsible manner'. He joined Sir Hugh Rossi in arguing that, unless action was taken without delay, damage in Britain and elsewhere would be irreversible and in the long term would cost far more to mitigate than the sums proposed in the Select Committee's report. Sir Hugh Rossi, in a forthright defence of the Select Committee's report, noted in the government's defence that it was understandable that the government was reluctant to undertake massive expenditure when it was preoccupied with finding every means of reducing public expenditure. He also outlined the government's concern when receiving advice from not disinterested sources that it was *not* certain that the proposed expenditure would solve the problem. The question of costs was raised in cavalier style by Willy Hamilton (Labour member for Fife Central), who pointed out that the £1,500 million cost of converting power stations was of a similar magnitude to what the government would be spending in the Falkland Islands over the next three years. The hapless William Waldegrave, forced once again to protect his political patrons by defending the government's position, said that the government took the subject and the Committee report very seriously, and accepted that the report had identified some gaps in research and monitoring that needed to be filled.

OFFICIAL ATTITUDES

Since the storm broke in December 1984 the British government has doggedly stuck to its position on acid rain. Scientific studies launched since 1982 (especially those funded under the Royal Society study) have been progressing at home and abroad, and the catalogue of reports on acid rain continues to grow. Unfortunately the debate rages on, mostly beyond the confines of Westminster. The conclusion drawn by Pearce Wright (science correspondence for *The Times*) in July 1984 that 'the controversy about the damage caused by acid rain is no closer to being resolved' (31 July 1984, p. 14) is still true over three years and many scientific reports later.

The government and the CEGB share a common line on acid rain, founded on various lines of evidence that either underline the inherent complexity of the issue or suggest that the sources and mechanisms of acid rain are *not* as previously thought. There are a great many pieces to the jigsaw, but the following repertoire of arguments is cited most regularly in defence by both the CEGB and the government:

(a) There is evidence that the acidity of rainfall in Britain has increased in recent years, although the complete picture has not yet emerged.
(b) Emissions of SO_2 in Britain (which come mainly from coal-fired power stations but also from factories) fell by 30 per cent between 1970 and 1982, and continue to fall. SO_2 emissions are now around the same level as they had been in the late 1940s.
(c) There appears to be a seasonal lag between peaks of oxide emissions and production of acid rain – the former occur in December and January, whereas rainfall acidity tends to be highest in May (when emissions are about two-thirds of peak levels).
(d) The links between production of acid rain and emissions of SO_2 and NO_x are difficult to establish, particularly given the evidence for (a)–(c) above.
(e) Because of (d), there are conflicting views on the likely impacts of reducing oxide emissions.
(f) Emissions of sulphur dioxide appear to be less important in producing acid rain than nitrogen oxides, for which local traffic may be more to blame than distant power stations. The government argues that reductions of oxide emissions from cars should be phased over a long period, and proposes that they should be based on gradual improvements with the adoption of lean burn engines rather than fitting catalytic burners to clean up exhaust gases.
(g) The chemistry of acid rain is now thought to be much more complex than previously believed – the significance of variables such as ozone,

sunlight, atmospheric oxygen, and moisture is as yet little understood.

(h) Patterns of damage attributed to acid rain appear to be extremely patchy, so that local factors (such as geology) may be highly significant. It is asked, for example, how acid rain can affect trees in West Germany and Switzerland, without – apparently – damaging trees in Norway, where concern centres on rivers and lakes.

(i) The evidence of acid rain damage in Britain (particularly the observed corrosion of buildings) may have developed incrementally over a longer period than the fifteen to twenty years often assumed, and the 'problem' is not of recent making. The CEGB, for example, attributes much of the observed damage to high levels of SO_2 that had existed in Britain's cities for much of this century, coming from millions of tons of coal burnt in a myriad of small urban sources (mainly domestic chimneys and factories) and in the type of municipal power station that has now long since been closed.

(j) It is still not possible definitely to attribute woodland damage in West Germany and elsewhere to acid rain – the links are inferred. There is a growing belief that ozone, rather than acid rain *per se*, may be responsible for the observed tree deaths. To mid-1985 no acute damage that could definitely be attributed to acid rain had been reported from British coniferous forests.

(k) The observed fish deaths (in Scandinavia, Wales, Scotland, and the Lake District) are probably caused by heavy metal poisoning, via natural aluminium dissolved from certain soils by the acid rain.

(l) Prevailing wind patterns make central European sources a potentially more significant contributor of the oxides that are blown over West Germany and Scandinavia than had previously been supposed.

(m) The estimated atmospheric budgets of sulphur and nitrogen over Europe have been altered on various occasions in recent years as new observations become available. Britain's alleged role as an exporter of oxides changes through time in harmony with the available data, and the picture is as yet far from properly established.

(n) The arguments outlined above underline the government's continued insistence that the case for solving the acid rain problem by reducing SO_2 emissions from coal-fired power stations is not yet proven. Given this belief, the need to conform to the EEC's proposed Directive is not yet established.

(o) Pollution control of the magnitude required to meet the Directive would require the installation of sulphur emission control equipment at ten of the CEGB's existing 2,000 MW coal-fired power stations by 1995. This would cost £1,500 million in capital investment (£150 million per station), and increased operating costs of around £250 million per annum. Electricity prices to consumers would need to rise by around 10 per cent over the next decade to pay for these extra production costs.

(p) The high costs of reducing emissions of SO_2 must be set alongside any benefits that might arise from reducing sulphur deposition by a few per cent. Although attempts to estimate costs and benefits are still as yet at the stage of educated guesses, the government argues that benefits are likely to be much smaller than costs (though this inevitably depends on what factors are taken into account, and how the costings are arrived at).

(q) The fact is also inescapable that many of the inferred benefits associated with reduced SO_2 emissions would be enjoyed by other countries downwind of Britain's power stations (especially Scandinavia and West Germany), whereas the costs would have to be met largely by British electricity consumers.

1986 AND ALL THAT

The European political campaign urging the UK to *act* on acid rain (chapter 8) built progressively from about 1982 onwards. Norway led the campaign with commendable determination and a keen sense of imminent victory. Rakel Surlien, the country's Minister of the Environment, commented on Britain's isolated position on acid rain during a three-day visit to London in mid-March 1986: 'I don't see how long you can remain isolated in this way. It must be very difficult to live with', she added poignantly (*The Times*, 17 March 1986, p. 3). Despite this cordial approach, there was to be no change in the Norwegian position, and no escape from the Scandinavian argument that Britain *must* join the '30 per cent club' (chapter 8) without further delay.

The Nordic 'war of attrition' was having some effect on the overtly arrogant attitude of the British government towards acid rain, and the frosty relations between Britain and Scandinavia were by now starting to thaw. A surprising switch in ground was apparent at the end of Mrs Surlien's visit to London, when William Waldegrave admitted – for the first time – that the British government accepted that sulphur emissions from British power stations were causing acid rain to fall on Norway (*The Times*, 20 March 1986, p. 9). 'We have no division as to the scientific perception of the problem', he added, although 'two or three years ago there was some scepticism in the scientific establishment'. Whilst there was now substantial agreement between Britain and Norway over the impact of acid rain on lakes and rivers, the two governments had not yet reached agreement on ways to reduce it.

The real motives behind this rather unexpected change in attitude by the British government may never emerge, but by now Britain was seen very much as a solitary island surrounded on its European coast by '30 per cent club' members. Moreover, acid rain had attracted widespread public interest via newspaper and television coverage, and environmental groups (like Friends of the Earth) had well co-ordinated campaigns against acid rain under way. Neither was the government pleased at the prospect of facing a biting tourist boycott during a year in which international tourism in Britain was otherwise depressed anyway (through unfavourable foreign exchange rates and the prospect of international terrorism). With both eyes firmly on the national political elections scheduled for 1987, the government was getting very conscious of its international image, its internal credibility, and its operational flexibility when faced with strong and ever-growing public pressure.

Norway, sensitive to this switch in attitude and keen to keep the issue alive, expressed hopes in April 1986 that Britain *would* join the '30 per cent club' after Prime Minister Margaret Thatcher visited Norway in September (*The Times*, 11 April 1986, p. 1). Early results from the country's RAIN (Reversing Acidification In Norway) project, which was scheduled to run from 1984 to 1989, were quickly published (*The Times*, 26 April 1986, p. 4) to encourage further this apparent slippage in Britain's hitherto unmoveable position. The RAIN project was organized by the Norwegian Institute for Water Research, and it involved fitting roofing over selected areas of forest (each one roughly the size of a tennis court). Some of the study areas – now effectively sealed from receiving atmospheric inputs – were irrigated with clean water, others with the acid rain that was falling over the surrounding forests. The early results showed as conclusively as anyone could have hoped that acid rain *was* causing forest damage. They were seized upon by the Scandinavians, who were keen to show how reducing SO_2 emissions would be an important step towards reducing if not reversing forest damage.

Further momentum in the campaign against Britain was gained in May, when the new Labour government was elected in Norway. Its manifesto had promised a more aggressive line on environmental issues, and the government wasted no time in winning its political colours. In mid-June Sissel Roenbell, the new minister for the environment, warned Britain that the 'softly, softly' approach on vexacious issues (like acid rain) taken by her Conservative predecessors would soon be a thing of the past (*The Times*, 18 June 1986, p. 11). On World Environment Day (17 June 1986) she accused the British government of 'provocation against international society', and she criticized its failure to install pollution-control equipment at the new Drax B coal-fired power station (Figure 10.4).

By late 1986, Nordic bitterness over Britain's continued refusal to join the '30 per cent club' had become strong and vocal. This annoyance, said Sir Hugh Rossi (Chairman of the House of Commons Select Committee on the Environment) on 10 September 1986, was 'understandable and highly regrettable in its effects on previously excellent relations with Norway' (*The Times*, 11 September 1986, p. 9). Since 1970, he added, sulphur emissions in Britain had already fallen significantly with the switch from coal to natural gas, conversion to nuclear power, and the rundown of some heavy industries (see Figure 1.5).

However, the British government was clearly undergoing a change of heart and a change of stance on acid rain. When Prime Minister Margaret Thatcher visited Oslo on 11 September, she announced that Britain was –

after all – intending to cut SO$_2$ emissions, but not by enough to qualify for entry to the '30 per cent club' (*The Times*, 11 September 1986, p. 9). The British government agreed to spend £600 million over the next ten years to curb British emissions of SO$_2$. The plan was to build FGD plants (see chapter 7) beside three of the country's twelve coal-fired power stations, at a cost of around £200 million each (*The Times*, 12 September 1986, p. 2). The stations are at Drax in North Yorkshire, West Burton in the West Midlands (Figure 10.4), and Fiddler's Ferry in Cheshire. It was predicted that up to 2,000 new jobs would be created in the control works, but fears were also expressed that the likely 1.5 per cent rise in electricity prices *could* lead to job losses elsewhere in the economy. This limited FGD programme was designed to reduce national SO$_2$ emissions by 14 per cent – a significant shortfall on the 30 per cent target being pursued in many countries (see Table 8.5). It was immediately dismissed as 'too little and too late' by the opposition Labour Party and by Friends of the Earth.

CONCLUSIONS

Britain has long occupied one of the hot seats in the acid rain debate. The country that had been tackling serious air pollution throughout the present century has, in the past decade, emerged as a sleeping giant who is reluctant to curb SO$_2$ emissions. Both the government and the Central Electricity Generating Board have argued that there are still too many fundamental research questions as yet unasked or unanswered, so that more research is needed 'before extremely costly programmes for reducing sulphur emissions are imposed without regard to whether they will be effective or not' (*The Times*, 2 April 1983, p. 7). Both parties have claimed to be ready and willing to act on acid rain, once the evidence justifies action.

The CEGB attitude is easy to understand, given that it has recurrently been singled out as the villain of the piece and that it would have to make the massive investments required to reduce SO$_2$ emissions to conform with the European Directive, and pass them on to its customers.

The British government's attitude towards acid rain has been entrenched, to say the least, and it has faced a barrage of opposition from home and abroad, from scientists (including government experts and advisers) and politicians (including the Lords and Commons Select Committees and many Tory MPs and MEPs), as well as the environmental lobby (such as Friends of the Earth). But it is perhaps wrong to think of a *monolithic* government attitude, shared by *all*. For example, the Tory MEPs who rebelled in Europe believed that they had the tacit support of Mr William Waldegrave and his chief, the Secretary of State for the Environment (Mr Patrick Jenkin), of the Foreign Secretary (Sir Geoffrey Howe), and of a majority of the Cabinet. Opposition within the Cabinet was believed to be limited to two key figures – the Secretary of State for Energy (Mr Peter Walker), who feared that anti-pollution measures would increase electricity costs, and the Prime Minister (Mrs Margaret Thatcher) who supported him (*The Times*, 7 December 1984, p. 5).

The real reasons underlying the government's attitudes to acid rain may never be revealed. The private thoughts of Cabinet ministers are not widely broadcast, especially during the life of their government. There are various key ingredients to the puzzle.

First, there is nothing unusual in a government of the day ignoring Select Committee reports. Indeed, the present Conservative government has a distinguished track record on this, having elected, for example, to ignore Select Committee recommendations *not* to abolish the metropolitan authorities in England and Wales.

Secondly, it is significant that the government has *not* really been faced with a hard-hitting campaign, pushing it relentlessly on the issue of acid rain. Much of the debate has been high level, with the environmental lobby contributing when directly prompted to do so by government announcements (which have doubtless been timed with fine political precision). The issue has *not* fully been taken up in an impassioned, well-orchestrated and widely publicized debate – partly because it is so complex, and because of the wide spectrum of views even amongst accredited specialists. The spectre of nuclear alternatives must also lurk in the minds of those who oppose the present government position on sulphur dioxide emissions.

Thirdly, as noted earlier, the 'case not proven' argument has been used repeatedly by the present government in its dealings over acid rain, as indeed it has by the Reagan administration in the USA (chapter 9). Ideology often takes precedence over argument in political decision making, and this is so particularly when politicians are faced with complex scientific arguments and unresolved scientific problems and uncertainties (as with acid rain or nuclear power).

Another element of the debate, which has remained largely hidden from the public gaze but has surfaced on a number of occasions, is Treasury pressure. The prospects of adding around £1,500 million to the costs of running the CEGB are not attractive to a government hot in pursuit of cutbacks in public expenditure, even if the costs can be passed on to consumers via electricity price rises. A Conservative government even more seduced than previous Tory governments by financial balance sheet approaches to planning has little time for what it sees as unnecessary expenditure. Such investment is even less welcomed by a government intent on wholesale privatization of previously nationalized industries. Privatization of British Telecom late in 1985 netted around £3.9 billion for the government; large sums were also raised from selling off British Gas late in 1986 and British Airways in early 1987. Should the Thatcher government envisage privatizing the nation's electricity industry, it would want to offer a 'cost-effective' and streamlined industry, unencumbered by costly anti-pollution requirements. The government must see it as imprudent to load the industry with extra costs and hence reduce its saleability on the cost-conscious open market.

There are other interpretations of the British government's extreme reluctance to act on acid rain. One is the possible desire to delay costly pollution control decisions until after the 1987 national elections. An alternative interpretation is that the government, which is openly committed to increasing the production of electricity by nuclear means, may seek to delay and frustrate attempts to 'clean up' coal-fired power stations until such time as the costs of expanding a nuclear programme become more viable. By that time also public opposition to SO_2 emissions might have reached significant levels. The government could then gallantly step in and seemingly solve the acid rain problem (and 'modernize') by opting for nuclear power capacity.

Yet another interpretation is that the government, faced with so many pressing problems with a higher political profile than acid rain – such as unemployment, public spending restraints, and continued opposition from trade unions – has little spare capacity on its political agenda to accommodate acid rain. The rationale of excluding items regarded to have minor political significance is *not* unknown to governments of all persuasions.

At the end of the day, the truth behind government reasoning will probably never be revealed. One thing, however, is clear. Acid rain is a real problem – even if the scientific arguments can be resolved, it will take a long time before the political debate is fully exhausted.

Bibliography

Abrahamsen, G. Bjor, K., Hornfeldt, R., and Tveite, B. (1976) *Effects of Acid Precipitation on Forests and Freshwater Ecosystems in Norway*, Oslo: SNSF Project.

Abrahamsen, G., Stuanes, A.O., and Tveite, B. (1983) 'Effects of long range transported air pollutants in Scandinavia', *Water Quality Bulletin* 8: 89–95, 109.

Abrahamsen, G. (1984) 'Effects of acidic deposition on forest soil and vegetation', *Philosophical Transactions of the Royal Society of London* 305: 369–82.

Ackerman, W. C., Harmeston, R.H., and Sinclair, R.A. (1970) 'Some long term trends in water quality of rivers and lakes', *Eos* 51: 516–22.

Agren, C. (1984a) '34 percent of West German forest land damaged', *Acid News* 1: 6–7.

Agren, C. (1984b) 'FRG: forest damage accelerating', *Acid News* 5: 12–13.

Air Conservation Commission (1965) *Air Conservation*. Washington, DC: American Association for Advancement of Science.

Air Pollution Control Association Panel (1980) 'Acid rain: an international concern', *Journal of the Air Pollution Control Association* 30: 1089.

Alabaster, J. S. and Lloyd, R. (1980) *Water Quality Criteria for Freshwater Fish* London: Butterworth.

Allar, B. (1984) 'Atmospheric fluidized bed combustion; no more coal-smoked skies?' *Environment*, March: 25–31.

Almer, B., Dickson, W., Ekstrom, C., Hornstrom, E., and Miller, U. (1974) 'Effects of acidification on Finnish lakes', *Ambio* 3: 30–6.

Anderson, J. G., Hazen, N.L., McLaren, B.E., Rowe, S.P., Schiller, C.M., Schwab, M.J., Solomon, L., Thompson, E.E., and Weinstock, E.M. (1985) 'Free radicals in the stratosphere; a new observational technique', *Science* 228: 1309–11.

Aniansson, B. (1984) 'A political breakthrough: but much work remains', *Acid Magazine* 1: 6–8.

Aniansson, B. (1985) 'A firm commitment', *Acid Magazine* 3: 29–30.

Anon (1976) 'Report from the international conference on the effects of acid precipitation in Telemark, Norway, June 14–19, 1976', *Ambio* 5: 201–2.

Anon (1981) 'Is all acid rain polluted?' *Science* 212: 1014.

Anon (1984a) 'Acid rain blights UK trees', *Acid News* 4: 10.

Anon (1984b) 'Twelve power stations must be cleaned up', *Acid News* 5: 6.

Anon (1984c) 'Correction', *Acid News* 5: 14.

Anon (1984d) 'How to neutralize acid rain', *Nature* 310: 611–12.

Anon (1984/5a) 'Reducing SO_2 emissions', *Acid Magazine* 2: 6–7.

Anon (1984/5b) 'Facing similar problems', *Acid Magazine* 2: 10–11.

Anon (1984/5c) 'Liming is the only chance', *Acid Magazine* 2: 15.

Anon (1984/5d) 'Corrosion from soil and water', *Acid Magazine* 2: 16.

Anon (1984/5e) 'Metals in foodstuffs and water', *Acid Magazine* 2: 17.

Anon (1984/5f) 'Signs of progress', *Acid Magazine* 2: 18–20.

Anon (1985) 'United Kingdom tree dieback: new survey', *Acid News* 4: 8.

Anon (1986a) 'West Germany: trend has not been halted', *Acid News* 1: 12–13.

Anon (1986b) 'Don't rain on Gro's parade', *The Economist* 13 September: 90–1.

Apling, A. (1981) *Air Pollution from Oxides of Nitrogen, Carbon and Hydrocarbons*, Stevenage: Warren Spring.

Arvidsson, L. (1985) 'A very special growth', *Acid Magazine* 3: 27–8.

Ashmore, M., Bell, N., and Rutter, J. (1985) 'The role of ozone in forest damage in West Germany', *Ambio* 14: 81–7.

Bache, B. W. (1985) 'Soil acidification and aluminium mobility', *Soil Use and Management* 1: 10–14.

Bache, B. W. (1986) 'Aluminium mobilization in soils and waters', *Journal of the Geological Society* 143: 699–706.

Ballance, R. C. and Olson, B. H. (1980) 'Water quality and health', pp. 173–92 in A. M. Gower (ed.) *Water Quality in Catchment Ecosystems*, London: Wiley.

Barabas, S. (1983) 'An acid test for acid precipitation', *Water Quality Bulletin* 8: 114.

Barker, M. L. (1974) 'Information and complexity – conceptualization of air pollution by specialist groups', *Environment and Behaviour* 6: 346–77.

Barnes, R. A. (1976) 'Long-term mean concentrations of atmospheric smoke and sulphur dioxide in country areas of England and Wales', *Atmospheric Environment* 10: 619–31.

Barratt, C. P. and Irwin, J. G. (eds) (1983) *Acid Deposition in the United Kingdom*, Stevenage: Warren Spring.

Barrett, E. and Brodin, G. (1955) 'The acidity of Scandinavian precipitation, *Tellus* 7: 251–7.

Barrett, M. and Dudley, N. (1984) *The Acid Rain Controversy*, London: Environmental Resources Research.

Barrie, L. M., Hoff, R. M., and Dagguptaty, S. M. (1981) 'The influence of mid-latitudinal pollution sources on haze in the Canadian arctic', *Atmospheric Environment* 15: 1407–19.

Battarbee, R. W. (1984) 'Diatom analysis and the acidification of lakes', *Philosophical Transactions of the Royal Soc. of London, Series B*, 305: 451–77.

Battarbee, R. W. and Flower, R. (1982) *Acidification of Galloway Lakes – Evidence from Diatoms*, London: University College.

Battarbee, R. W., Appleby, P. G., Odell, K., and Flower, R. J. (1985) '210 Pb dating of Scottish lake sediments, afforestation and accelerated soil erosion', *Earth Surface Processes and Landforms* 10: 137–42.

Battarbee, R. W., Flower, R. J., Stevenson, A. C., and Rippey, B. (1985a) 'Lake acidification in Galloway – a palaeoecological test of competing hypotheses', *Nature* 314: 350–2.

Battarbee, R. W., Flower, R. J., Stevenson, A. C., and Rippey, B. (1985b) 'What evidence is there for acid rain affecting freshwater ecosystems in the past?' pp. 89–95 in Scottish Wildlife Trust, *Report of the Acid Rain Inquiry, 27–29 September 1984*, Edinburgh: Scottish Wildlife Trust.

Beamish, R. J. and Harvey, H. H. (1972) 'Acidification of the La Cloche Mountain Lakes, Ontario, and resulting fish mortalities', *Journal of the Fisheries Research Board of Canada* 29: 1131–43.

Beamish, R. J., Lockart, W. L., van Loon, T. E., and Harvey, H. (1975) 'Long term acidification of a lake and resulting effects on fishes', *Ambio* 4: 98–102.

Beaumont, J. *et al.* (eds) (1984) *The Ecological Effects of Deposited Sulphur and Nitrogen Compounds*, London: Royal Society.

Beaver Committee (1954) *Final Report of the Beaver Committee*, Cmd 9322, London: HMSO.

Bell, J. N. B. and Clough, W. S. (1973) 'Depression of yield in ryegrass exposed to sulphur dioxide', *Nature* 241: 47–9.

Benarie, M. (1982) 'The influence of emission height on the meso-scale and long-range transport of reactive pollutants', *Science of the Total Environment* 23: 163–72.

Benarie, M. (1985) 'Lack of relationship between acidic rain and tree damage in urban areas of Europe', *Science of the Total Environment* 43: 185–6.

Bengtsson, B., Dickson, W., and Nyberg, P. (1980) 'Liming acid lakes in Sweden', *Ambio* 9: 34–6.

Bennett, B. G., Kretzschmar, J.G., Akland, G.A., and de Konig, H.W. (1985) 'Urban air pollution worldwide', *Environmental Science and Technology* 19: 298–305.

Berlekom, M. (1985a) 'Birds of the lake', *Acid Magazine* 3: 6–9.

Berlekom, M. (1985b) 'Frogs', *Acid Magazine* 3: 11–12.

Berlekom, M. (1985c) 'First to suffer', *Acid Magazine* 3: 13–14.

Berlekom, M. (1985d) 'Sensitive to polluted air', *Acid Magazine* 3: 25–6.

Bhumralker, C. M. (ed.) (1984) *Meteorological Aspects of Acid Rain*, London: Butterworth.

Binns, W. O. (1984a) *Acid Rain and Forestry*, Edinburgh: Forestry Commission.

Binns, W. O. (1984b) 'Acid rain: farming and forestry', *Coal Energy Quarterly* 41: 2–13.

Binns, W. O. *et al.* (1985) *Forest Health and Air Pollution: 1984 Survey*, Edinburgh: Forestry Commission.

Blank, L. W. (1985) 'A new type of forest decline in Germany', *Nature* 314: 311–14.

Bobee, B. and Lachance, M. (1983) 'A method of analyzing data on lake acidification and its application in Quebec, Canada', *Water Quality Bulletin* 8: 160–7.

Bolin, B. and Charlson, R. J. (1976) 'On the role of the tropospheric sulphur cycle in the shortwave radiative climate of the earth', *Ambio* 5: 47–54.

Bormann, F. H. and Likens, G. E. (1979) *Pattern and Process in a Forested Ecosystem*, New York: Springer-Verlag.

Bottam, L. J. (1966) *The Unclean Sky*, New York: Doubleday.

Brady, G. L. and Selle, J. C. (1985) 'Acid rain – the international response', *International Journal of Environmental Studies* 24: 217–30.

Brady, M. (1985) 'The relation of acid rain to energy policy', *Energy* 10: 1113–8.

Braekke, F. H. (1976) *Impact of Acid Precipitation on Forests and Freshwater Ecosystems in Norway*, Oslo: SNSF Project.

Bricker, O. P. (ed.) (1984) *Geological Aspects of Acid Deposition*, London: Butterworth.

Bricker, O. P. (1986) 'Geochemical investigations of selected eastern United States watersheds affected by acid deposition', *Journal of the Geological Society* 143: 621–6.

Briggs, A. (1980) *Victorian Cities*, Harmondsworth: Penguin.

Brimblecombe, P. (1977) 'London air pollution 1500–1900', *Atmospheric Environment* 11: 1157–62.

Brimblecombe, P. (1985) 'Is acid rain a new phenomenon?' pp. 5–12 in Scottish Wildlife Trust, *Report of the Acid Inquiry, 27–29 September 1984*, Edinburgh: Scottish Wildlife Trust.

Brimblecombe, P. and Stedman, D. (1982) 'Historical evidence for a dramatic increase in the nitrate component of acid rain', *Nature* 298: 460.

Brown, D. J. A. (1983) 'Effects of calcium and aluminium concentrations on the survival of brown trout (*Salmo trutta*) at low pH', *Bulletin of Environmental Contamination and Toxicology* 30: 582–7.

Brown, K. (1982) 'Sulphur in the environment', *Environmental Pollution* 3: 47–80.

Bryce-Smith, D. (1985) 'Acid rain', pp. 2–3 in *Acid Rain: Report of the One Day Conference Held at Shire Hall, Reading (4 March 1985)*, London: National Council of Women of Great Britain.

Buckley-Golder, D. H. (1984) *Acidity in the Environment*, Harwell: Energy Technology Support Unit.

Buckley-Golder, D. (1985) 'Acidity in the environment', pp. 3–5 in *Acid Rain: Report of the One Day Conference Held at Shire Hall, Reading (4 March 1985)*, London: National Council of Women of Great Britain.

Buckner, M. (1985) 'Impacts of proposed acid rain legislation on employment', *Journal of the Air Pollution Control Association* 35: 221–3.

Budiansky, S. (1984) 'Canada must act alone', *Nature* 307: 679.

Bunyard, P. (1979) 'Living on a knife-cdge: the aftermath of Harrisburg', *The Ecologist* 9: 97–102.

Burroughs, W. J. (1981) 'Mount St Helens – a review', *Weather* 36: 238–40.

Burt, T. P. and Day, M. R. (1977) 'Spatial variations in rainfall and stream water quality around Avonmouth industrial complex', *International Journal of Environmental Studies* 11: 205–9.

Butler, J. D. (1979) *Air Pollution Chemistry*, New York: Academic Press.

Cadle, S. H., Dasch, J. M., and Grassnickla, N. E. (1984) 'Northern Michigan snowpack – a study of acid stability and release', *Atmospheric Environment* 18: 807–16.

Calvert, J. G. (ed.) (1984) SO_2, NO_x and NO_2 Oxidation Mechanisms; Atmospheric Considerations, London: Butterworth.

Calvert, J. G., Lazrus, A., Kok, G. L., Heikes, B. G., Walega, J. G., Lind, J., and Cantrell, C. A. (1985) 'Chemical mechanisms of acid generation in the atmosphere: review article', *Nature* 317: 27–35.

Camuffo, D., Moule, M. D., and Ongaro, A. (1984) 'The pH of the atmospheric precipitation in Venice, related to both the dynamics of precipitation events and the weathering of monuments', *Science of the Total Environment* 40: 125–39.

Carrick, T. R. (1979) 'The effects of acid water on the hatching of salmonid eggs', *Journal of Fish Biology* 14: 165–72.

Carr, D. E. (1965) *The Breath of Life*, New York: Norton.

Carson, R. (1962) *Silent Spring*, New York: Hamilton.

Carter, F. W. (1985) 'Pollution problems in post-war Czechoslovakia', *Transactions of the Institute of British Geographers* 10: 17–44.

Central Electricity Generating Board (1979) *Effects of Sulphur Dioxide and its Derivatives on Health and Ecology*, Leatherhead: CEGB.

Central Electricity Generating Board (1984) *Acid Rain*, London: CEGB.

Chan, W. H., Ro, C. U, Lusis, M. A., and Vet, R. J. (1982) 'Impact of the Inco nickel smelter emissions on precipitation quality in the Sudbury area', *Atmospheric Environment* 16: 801–14.

Chan, W. H., Vet, R. J., Ro, C. U., Tang, A. J. S., and Lusis, M. A. (1984) 'Impact of Inco smelter emissions on wet and dry depositions in the Sudbury area', *Atmospheric Environment* 18: 1001–8.

Charlson, R. J. and Rodhe, H. (1982) 'Factors controlling the acidity of natural rainwater', *Nature* 295: 683–5.

Charlson, R. J. *et al.* (1974) 'Background aerosol: optical detection in the St Louis region', *Atmospheric Environment* 8: 1257–67.

Clapham, W. B. (1973) *Natural Ecosystems*, New York: Macmillan.

Clark, P. A., Fletcher, J. S., Kallend, A. S., McElroy, W. J., Marsh, A. R. W., and Webb, A. H. (1984) 'Observations of cloud chemistry during long range transport of power plant plumes', *Atmospheric Environment* 18: 1849–58.

Clarke, A. J. (1985) 'Control measures', pp. 5–6 in *Acid Rain; Report of the One Day Conference Held at Shire Hall, Reading, on 4 March 1985*, London: National Council of Women of Great Britain.

Clayton, K. (1971) 'Reality in conservation', *Geographical Magazine* 44: 66–7.

Cocks, A. T., Kallend, A. S., and Marsh, A. R. W. (1983) 'Dispersion limitations of oxidation in power plant plumes during long-range transport', *Nature* 305: 122–3.

Cogbill, C. V. and Likens, G. E. (1974) 'Acid precipitation in the north east United States', *Water Resources Research* 10: 1133–7.

Colbeck, I. (1985) 'Photochemical ozone pollution', *Weather* 40: 241–4.

Collins (1982) *Collins Dictionary of the English Language*, Glasgow: Collins.

Commission of the European Communities (1970) 'Council Directive on the approximation of the laws of member states relating to measures to be taken against air pollution by gases from positive ignition engines of motor vehicles', 70/220/EEC, *Official Journal of the European Community* L76.1.

Commission of the European Communities (1980) 'Council Directive on air quality limit values and guide values for sulphur dioxide and suspended particulates', 80/779/EEC, *Official Journal of the European Community* 1229: 30–48.

Commission of the European Communities (1983) 'Commission proposal for a Council Directive on air quality for nitrogen dioxide', *Official Journal of the European Community* C258: 3–7.

Commission of the European Communities (1984a) 'Proposal for the amendment of the EEC Directives on the lead content of petrol and motor vehicle emission', COM(84)226 final, *Official Journal of the European Communities* C178: 5–8.

Commission of the European Communities (1984b) 'Council Directive on the combating of air pollution from industrial plant', 84/360/EEC, *Official Journal of the European Community* L188: 20–5.

Commission of the European Communities (1984c) 'Proposal for a Council Directive on the limitations of emissions of pollutants into the air from large combustion plants', COM(38)704 final, *Official Journal of the European Communities* C49: 1–7.

Commission of the European Communities (1985) *Report on the Action of the Commission of the European Economic Community on Acid Deposition*, Luxembourg: CEC.

Commoner, B. (1985) 'Economic growth and environmental quality: how to have both', *Social Policy* 16: 18–26.

Courage, I. H. (1983) 'Control of acid precipitation by public international law', *Water Quality Bulletin* 8: 133–6.

Courtis, M. W. (1984) 'Nitrogen dioxide: is there a problem?' *Environmental Health* 92: 268–71.

Cowling, E. B. (1982) 'Acid precipitation in historical perspective', *Environmental Science and Technology* 16: 110–23.

Cowling, E. B., Jones, L. H. P., and Lockyer, D. R. (1973) 'Increased yield through correction of sulphur deficiency in ryegrass exposed to sulphur dioxide', *Nature* 243: 479–80.

Crawshaw, D. H. (1986) 'The effects of acid runoff on streams in Cumbria', pp. 25–31 in P. Ineson (ed.) *Pollution in Cumbria*, Grange over Sands: ITE.

Cresser, M. S., Edwards, A. C., Ingham, S., Skiba, V., and Pierson-Smith, T. (1986) 'Soil–acid deposition interactions and their possible effects on geo-chemical weathering rates in British uplands', *Journal of the Geological Society* 143: 649–58.

Crocker, T. D. (ed.) (1984) *Economic Perspectives on Acid Deposition Control*, London: Butterworth.

Crocker, T. D. and Regens, J. L. (1985) 'Acid deposition control: a cost benefit analysis', *Environmental Science and Technology* 19: 112–16.

Cronan, C. S. and Schofield, C. L. (1979) 'Aluminium leaching response to acid precipitation: effects on high elevation watersheds in the north east', *Science* 204: 304–6.

Crowe, J. (1968) 'Towards a "definitional model" of public perceptions of air pollution', *Journal of the Air Pollution Control Association* 18: 154–8.

Cullis, C. F. and Hischler, M. M. (1980) 'Atmospheric sulphur: natural and man-made sources', *Atmospheric Environment* 14: 1263–78.

Davies, G. L. (1968) *The Earth in Decay*, London: Macdonald.

Davies, T. (1980) 'Grasses more sensitive to SO_2 pollution in conditions of low irradience and short days', *Nature* 284: 483–5.

Davies, T. D., Abrahams, P. W., Tranter, M., Blackwood, I., Brimblecombe, P., and Vincent, C. E. (1984) 'Black acidic snow in the remote Scottish Highlands', *Nature* 312: 58–61.

Davis, R. and Berge, F. (1980) 'Atmospheric deposition in Norway during the last 300 years as recorded in SNSF lake sediments, II. Diatom stratigraphy and inferred pH', pp. 270–1 in D. Drablos and A. Tollan (eds) *Ecological Impacts of Acid Precipitation*, Oslo: SNSF Project.

Davison, A. W. (1984) 'Ecological assessment of the effects of atmospheric emissions', pp. 376–86 in R. D. Roberts and T. M. Roberts (eds) *Planning and Ecology*, London: Chapman & Hall.

Dear, C. and Laird, C. (1984) 'In pursuit of acid rain', *New Scientist* 104: 30–3.

Deevey, E. S. (1970) 'Mineral cycles', *Scientific American* September: 148–58.

Department of Environment (1976) *Effects of Airborne Sulphur Compounds on Forests and Freshwaters*, London: HMSO.

Department of Environment (1983) *Digest of Environmental Protection and Water Statistics*, London: HMSO.

Department of Environment (1984) *Environmental Protection: Problems, Progress, Practice, Principles and Prospects*, London: HMSO.

Dicks, B. (1985) 'Cloud over Athens', *Geographical Magazine* 57: 298–9.

Dickson, W. (1975) *Acidification of Swedish Lakes*, Report 54, Drottningholm, Sweden: Institute of Freshwater Research.

Dickson, W. (1978) 'Some effects of the acidification of Swedish lakes', *Verhandlungen der Internationalen Vereinigung für Theoretische und Angewandte Limnologie* 20: 851–6.

Diprose, G. and Pumphrey, N. W. J. (1984) 'Assessing the impact of industrial emissions to the atmosphere', pp. 364–76 in R. D. Roberts and T. M. Roberts (eds) *Planning and Ecology*, London: Chapman & Hall.

Dochinger, L. S. and Seliga, T. A. (eds) (1976) *Proceedings of the First International Symposium on Acid Precipitation and the Forest Ecosystem*, Upper Darby, Pennsylvania: United States Department of Agriculture, N.E. Forest Experimental Station.

Dollard, G. J., Unsworth, M. H., and Harvey, M. J. (1983) 'Pollutant transfer in upland regions by occult precipitation', *Nature* 302: 241–3.

Douglas, I. (1983) *The Urban Environment*, London: Arnold.

Dovland, H. and Saltbones, J. (1979) *Emissions of Sulphur Dioxide in Europe, 1978*, Report EMEP/CCC 2/79, Oslo: Norwegian Institute for Air Research.

Drablos, D. and Tollan, A. (eds.) (1980) *Ecological Impacts of Acid Precipitation*, Norway: SNSF Project.

Driscoll, C. T., Baker, J. P., Bisogni, B. J., and Schofield, C. L. (1980) 'Effects of aluminium speciation on fish in dilute acidified waters', *Nature* 284: 161–4.

Dudley, N., Barrett, M., and Baldock, D. (1985) *The Acid Rain Controversy*, London: Earth Resources Research.

Duffus, J. (1980) *Environmental Toxicology*, London: Arnold.

Dunmore, J. (ed.) (1986) *NSCA Reference Book*, London: National Society for Clean Air.

Durham, J. L. (ed.) (1984) *Chemistry of Particles, Fogs and Rain*, London: Butterworth.

Durzel, R. B. and Cetrulo, C. L. (1981) 'The effects of environmental pollutants on human reproduction, including birth defects, *Environmental Science and Technology* 15: 626–39.

Dutt, P. (1984) 'Delhi gasps for its air', *Mazingira* 8: 10–11.

Edmunds, W. M. and Kinniburgh, D. G. (1986) 'The susceptibility of UK groundwaters to acidic deposition', *Journal of the Geological Society* 143: 707–20.

Egner, H. and Eriksson, E. (1955) 'Current data on the chemical composition of air and precipitation', *Tellus* 7: 134–9.

Eisenbud, M. (1970) 'Environmental protection in the city of New York', *Science* 170: 706–12.

Elder, F. C. (1983) 'Acid precipitation: an unseen plague of the industrial age', *Water Quality Bulletin* 8: 58.

Elgmore, K. *et al.* (1973) 'Polluted snow in southern Norway during the winters 1968–1971', *Environmental Pollution* 4: 41–52.

Eliassen, A. and Saltbones, J. (1983) 'Modelling of long-range transport of sulphur over Europe: a two year model run and some model experiments', *Atmospheric Environment* 17: 1457–73.

Elkins, C. L. (1985) 'Acid rain policy development', *Journal of the Air Pollution Control Association* 35: 202–4.

Elsom, D. (1984) 'Los Angeles smog threatens 1984 Olympic Games', *Weather* 39: 200–7.

Elsworth, S. (1984a) *Acid Rain*, London: Pluto Press.

Elsworth, S. (1984b) 'Acid rain in the UK', *Acid News* 3: 3–5.

Emberlin, J. C. (1980a) 'Smoke and sulphur dioxide in a rural area of central southern England', *Atmospheric Environment* 14: 1371–80.

Emberlin, J. C. (1980b) 'Smoke and sulphur dioxide concentrations in relation to topography in a rural area of central southern England', *Atmospheric Environment* 14: 1381–90.

Environmental Resources Ltd (1984) *Acid Rain – A Review of the Phenomenon in the EEC and Europe*, London: Graham & Trotman.

Evans, G. W. and Jacobs, S. V. (1981) 'Air pollution and human behaviour', *Journal of Social Issues* 37: 95–125.

Evans, L. S. (1982) 'Biological effects of acidity in precipitation on vegetation: a review', *Environmental and Experimental Botany* 22: 155–69.

Fassina, V. (1978) 'A survey of air pollution and deterioration of stonework in Venice', *Atmospheric Environment* 12: 2205–11.

Faust, S. D. and Hunter, J. V. (eds) (1976) *Principles and Applications of Water Chemistry*, New York: Wiley.

Fenton, A. F. (1960) 'Lichens as indicators of atmospheric pollution', *The Irish Naturalists' Journal* 13: 153–9.

Ferguson, P. and Lee, J. A. (1983) 'Past and present sulphur pollution in southern Pennines', *Atmospheric Environment* 17: 1131–7.

Ferry, B. W., Baddeley, M. S., and Hawksworth, D. L. (1973) *Air Pollution and Lichens*, London: Athlone Press.

Firket, J. (1936) 'Fog along the Meuse valley', *Transactions of the Faraday Society* 32: 1192–7.

Fisher, B. (1981) 'Acid rain – the long range transport of air pollutants', *Weather* 36: 367–9.

Fisher, B. (1982) 'Deposition of sulphur and the acidity of precipitation over Ireland', *Atmospheric Environment* 16: 21–43.

Fishlock, D. and Wilkinson, M. (1986) 'The make or break year', *The Financial Times* 25 February: 27.

Flower, R. J. (1984) Personal communication, reported in Greater London Council, *Acid Rain and London*, London: GLC, 1985.

Flower, R. J. and Battarbee, R. W. (1983) 'Diatom evidence for recent acidification of two Scottish lakes', *Nature* 305: 130–3.

Forestry Commission (1985) *Air Pollution and Tree Health*, Edinburgh: Forestry Commission.

Fowler, D., Cape, J. N., Leith, I. D., Paterson, I. S., Kinnaird, J. W., and Nicholson, I. A. (1982) 'Rainfall acidity in northern Britain', *Nature* 297: 383–6.

Fowler, D., Cape, J. N., Leith, I. D., and Paterson, I. S. (1986) 'Acid deposition in Cumbria', pp. 7–10 in P. Ineson (ed.) *Pollution in Cumbria*, Grange over Sands: ITE.

Francis, A. J. (1982) 'Effects of acid precipitation and acidity on soil microbial processes', *Water, Air and Soil Pollution* 18: 375–94.

Frenzel, G. (1985) 'The restoration of medieval stained glass', *Scientific American* 252: 100–6.

Friend, J. P. (1973) 'The global sulfur cycle', pp. 177–201 in S. I. Rasool (ed.) *Chemistry of the Lower Atmosphere*, New York: Plenum.

Friends of the Earth (1985) *Acid Rain*, London: FoE.

Frohlinger, J. O. and Kane, R. (1975) 'Precipitation: its acidic nature', *Science* 189: 455–7.

Fromm, P. D. (1980) 'A review of some physiological and toxicological responses of freshwater fish to acid stress', *Environment Biology and Fish* 5: 79–93.

Fry, G. L. A. and Cooke, A. S. (1984) *Acid Deposition and its Implications for Nature Conservation in Britain*, Shrewsbury: Nature Conservancy Council.

Galloway, J. N. and Whelpdale, D. M. (1980) 'An atmospheric sulphur budget for eastern North America', *Atmospheric Environment* 14: 409–17.

Galloway, J. N., Likens, G. E., and Edgerton, E. S. (1976) 'Acid precipitation in the north eastern United States: pH and acidity', *Science* 194: 722–3.

Galloway, J. N., Likens, G. E., and Hawley, M. E. (1984) 'Acid precipitation: natural versus anthropogenic components', *Science* 226: 829–30.

Galloway, J. N., Likens, G. E., Keene, W. C., and Miller, J. C. (1982) 'The composition of precipitation in remote areas of the world', *Journal of Geophysical Research* 87: 8771–86.

Garland, J. A. (1978) 'Dry and wet removal of sulphur from the atmosphere', *Atmospheric Environment* 12: 349–62.

Gates, D. M. (1972) *Man and his Environment: Climate*, London: Harper & Row.
Georgii, H. W. (1981) 'Review of the acidity of precipitation according to the WMO network', *Idojaras* 85: 1–9.
Gilbert, A. H. (1982) 'The economics of acid rain: an invisible hand of control', *International Journal of Environmental Studies* 18: 85–90.
Gilbert, O. L. (1970) 'Further studies on the effect of sulphur dioxide on lichens and bryophytes', *New Phytologist* 69: 605–27.
Gilbert, O. L. (1975) 'Effects of air pollution on landscape and land use around Norwegian aluminium smelters', *Environmental Pollution* 8: 113–21.
Gillette, G. D. (1975) 'Sulphur dioxide and material damage', *Journal of the Air Pollution Control Association* 25: 1238–43.
Glass, N. R., Arnold, D. E., Galloway, J. N., Hendrey, G. R., Lee, J. J., McFee, W. M., Norton, S. A., Powers, C. F., Rambo, D. L., and Schofield, C. L. (1982) 'Effects of acid precipitation', *Environmental Science and Technology* 16: 162A–169A.
Goldsmith, F. B. and Wood, J. B. (1983) 'Ecological effects of upland afforestation', pp. 287–311 in A. Warren and F. B. Goldsmith (eds) *Conservation in Perspective*, London: Wiley.
Gomez, B. and Smith, C. G. (1984) 'Atmospheric pollution and fog frequency in Oxford, 1926–1980', *Weather* 39: 379–84.
Gorham, E. (1955) 'On the acidity and salinity of rain', *Geochimica et Cosmochimica Acta* 7: 231–9.
Gorham, E. (1958) 'The influence and importance of daily weather conditions in the supply of chloride, sulphate and other ions to fresh water from atmospheric precipitation', *Philosophical Transactions of the Royal Society of London, Series B* 241: 147–78.
Gorham, E. (1982) 'Robert Angus Smith, FRS, and "chemical climatology"', *Notes and Records of the Royal Society of London* 36: 267–72.
Goudie, A. (1981) *The Human Impact*, Oxford: Blackwell.
Goudie, A. (1984) *The Nature of the Environment*, Oxford: Blackwell.
Gower, A. M. (1980) 'Ecological effects of changes in water quality', pp. 145–72 in A. M. Gower (ed.) *Water Quality in Catchment Ecosystems*, London: Wiley.
Grahn, O. (1980) 'Fish kills in two moderately acid lakes due to high aluminium concentration', in D. Drablos and A. Tollan (eds) *Ecological Impacts of Acid Precipitation*, Oslo: SNSF.
Granat, L. (1977) 'Sulphate in precipitation as observed by the European Atmospheric Chemistry Network', *Atmospheric Environment* 12: 413–24.
Grant, R. J. (1985) 'Evaluation of current legislative proposals for emission reductions', *Journal of the Air Pollution Control Association* 35: 216–18.
Greater London Council (1972) *Annual Report of the Scientific Advisor to the Greater London Council*, London: GLC.
Greater London Council (1973) *Annual Report of the Scientific Advisor to the Greater London Council*, London: GLC.
Greater London Council (1982) *A Review of Air Pollution by Sulphur Dioxide and Smoke in London*, London: GLC.
Greater London Council (1983) *Thirty Years on – A Review of Air Pollution in London*, London: GLC.
Greater London Council (1985) *Acid Rain and London*, London: GLC.

Hagen, A. and Langeland, L. (1973) 'Polluted snow in southern Norway and the effect of the meltwater on freshwater and aquatic organisms', *Environmental Pollution* 5: 200–5.
Hagstrom, T. (1980) 'Reproductive strategy and success in amphibians in waters

acidified by atmospheric pollution', *Proceedings of the European Herpetological Symposium*, Oxford.

Haines, T. A. (1981) 'Acidic precipitation and its consequences for aquatic ecosystems: a review', *Transactions of the American Fisheries Society* 110: 669–707.

Halberstadt, M. L. and Shiller, J. W. (1985) 'View of the motor vehicle industry on control of nitrogen oxide emissions', *Journal of the Air Pollution Control Association* 35: 219–21.

Hale, M. E. (1974) *The Biology of Lichens*, London: Arnold.

Halford, J. (1986) 'Germany's forests are dying', *Plain Truth* 51: 15–16, 24.

Hansen, D. and Hidy, G. (1982) 'Review of questions regarding rain acidity data', *Atmospheric Environment* 16: 2107–26.

Harriman, R. (1978) 'Nutrient leaching from fertilized forest watersheds in Scotland', *Journal of Applied Ecology* 15: 933–42.

Harriman, R. and Morrison, B. (1981) 'Forestry, fisheries and acid rain in Scotland', *Scottish Forestry* 35: 89–95.

Harriman, R. and Morrison, B. (1982) 'Ecology of streams draining forested and non-forested catchments in an area of central Scotland subject to acid precipitation', *Hydrobiologia* 88: 251–63.

Harriman, R. and Wells, D. (1985) 'Causes and effects of surface water acidification in Scotland', *Journal of the Institution of Water Pollution Control* 84: 215–24.

Harris, G. R. (1985) 'Environmentalism and politics in the USA; the historical underpinnings of hazardous waste management – a viewpoint', *International Journal of Environmental Studies* 24: 169–85.

Harvie, C., Martin, G., and Scharf, A. (eds) (1981) *Industrialization and Culture, 1830–1914*, London: Macmillan.

Havas, M. and Hutchinson, T. C. (1983) 'The Smoking Hills: natural acidification of an aquatic ecosystem', *Nature* 301: 23–7.

Hawkins, D. G. (1985) 'The benefits and costs of acid rain control' *Journal of the Air Pollution Control Association* 35: 224–5.

Hawksworth, D. L. (1971) 'Lichens as litmus for air pollution: a historical review', *International Journal of Environmental Studies* 1: 281–96.

Hawksworth, D. L. and Rose, C. I. (1981) 'Lichen recolonization in London's cleaner air, *Nature* 289: 289–92.

Hawksworth, D. L. and Rose, F., (1970) 'Qualitative scale for estimating sulphur dioxide air pollution in England and Wales using epiphytic lichens', *Nature* 227: 145–8.

Heath, J. (1982) *Threatened Butterflies in Europe*, Strasbourg: Council of Europe.

Heath, M. (1986) 'Polluted air falls on Spain', *New Scientist* 1526: 60–2.

Hegg, P. V. and Hegg, D. A. (1978) 'Oxidation of sulphur dioxide in aqueous systems, with particular reference to the atmosphere', *Atmospheric Environment* 12: 241–54.

Hendrey, G. R. (ed.) (1984) *Early Biotic Responses to Advancing Lake Acidification*, London: Butterworth.

Hendrey, G. R., Baalstrud, K., Traaen, T. S., Laake, M., and Roddum, G. (1976) 'Acid precipitation: some hydrobiological changes', *Ambio* 5: 224–7.

Henriksen, A. (1979) 'A simple approach for identifying and measuring acidification of fresh waters', *Nature* 278: 542–5.

Henriksen, A. and Kirkhusmo, L. A. (1982) 'Acidification of groundwater in Norway', *Nordic Hydrology* 13: 183–92.

Hicks, B. B. (ed.) (1984) *Deposition, Both Wet and Dry*, London: Butterworth.

Highton, N. H. and Chadwick, M. J. (1982) 'The effects of changing patterns of energy use on sulphur emissions', *Ambio* 11: 324–9.

Hileman, B. (1981) 'Acid precipitation', *Environmental Science and Technology* 15: 1119–24.

Hileman, B. (1984) 'Acid rain perspectives – a tale of two countries', *Environmental Science and Technology* 18: 341–4.

Hjelm, A. (1984) 'Liming the soil can result in poorer growth', *Acid Magazine* 1: 27–8.

HMSO (1984) *Acid Rain*, Cmnd 939, London: HMSO.

Holdgate, M. W. (1979) *A Perspective on Environmental Pollution*, Cambridge: Cambridge University Press.

Holdgate, M. W., Kassas, M., and White, G. F. (1982) *The World Environment, 1972–1982*, Dublin: Tycooly International.

Holtby, H. G. (1973) 'Air quality in a subarctic community', *Arctic* 26: 292–302.

Hooper, R. P. and Shoemaker, C. A. (1985) 'Aluminium mobilization in an acidic headwater stream: temporal variation and mineral dissolution disequilibria', *Science* 229: 463–5.

Hornung, M. (1985) 'Acidification of soils by trees and forests', *Soil Use and Management* 1: 24–8.

Hornung, M., Adamson, J. K., Reynolds, B., and Stevens, P. A. (1986) 'Influence of mineral weathering and catchment hydrology on drainage water chemistry in three upland sites in England and Wales', *Journal of the Geological Society* 143: 627–34.

House of Commons Environment Committee (1984) *Acid Rain*, London: HMSO.

House of Lords Select Committee on the European Communities (1984) *Air Pollution*, London: HMSO.

Howells, G. D. (1984) 'Fishery decline: mechanisms and predictions', *Proceedings of the Royal Society* 305B: 529–47.

Hoyle, R. (1982) 'The silent scourge', *Time* 8 November: 38–44.

Hultberg, H. and Johansson, S. (1981) 'Acid groundwater', *Nordic Hydrology* 12: 51–64.

Hutchinson, T. C. and Havas, M. (eds) (1980) *Effects of Acid Precipitation on Terrestial Ecosystems*, New York: Plenum.

Hutton, J. (1795) *Theory of the Earth: with Proofs and Illustrations*, Edinburgh.

Jacobsen, J. S. (1984) 'Effects of acid aerosol, fog, mist and rain on crops and trees', *Philosophical Transactions of the Royal Society of London* 305: 327–38.

Jacobsen, J. S. and Troiano, J. J. (1983) 'Dose–response functions for effects of acidic precipitation on vegetation', *Water Quality Bulletin* 8: 67–71, 109.

Jensen, K. W. and Snekvik, E. (1972) 'Low pH levels wipe out salmon and trout populations in southernmost Norway', *Ambio* 1: 223–5.

Jickells, T. A., Knap, A., Church, T., Galloway, J., and Miller, J. (1982) 'Acid rain in Bermuda', *Nature* 297: 55–7.

Johannessen, M. and Henriksen, A. (1978) 'Chemistry of snow meltwater: changes in concentration during melting', *Water Resources Research* 14: 615–19.

Johnson, B. (1976) 'International environmental conventions', *Ambio* 5: 55–65.

Johnson, N. M. (1979) 'Acid rain: neutralization within the Hubbard Brook ecosystem and regional implications', *Science* 204: 497–9.

Johnson, N. M., Driscoll, C. T., Eaton, J. S., Likens, G. E., and McDowell, W. M. (1981) 'Acid rain, dissolved aluminium and chemical weathering at the Hubbard Brook Experimental Forest, New Hampshire', *Geochimica et Cosmochimica Acta* 45: 1421–37.

Junge, C. E. (1974) 'Residence time and variability of tropospheric trace gases', *Tellus* 24: 477–88.

Kallend, A. S., Marsh, A. R. W., Pickles, J. H., and Proctor, M. V. (1983) 'Acidity of rain in Europe', *Atmospheric Environment* 17: 127–37.

Kasina, S. (1980) 'On precipitation acidity in south east Poland', *Atmospheric Environment* 14: 1217–21.

Kerch, R. L. (1985) 'Policy options for dealing with acid deposition: a high sulfur coal producer's perspective', *Journal of the Air Pollution Control Association* 35: 213–16.

Kerr, R. A. (1982) 'Tracing sources of acid rain causes big stir', *Science* 215: 881.

Killham, K., Firestone, M. K., and McGoll, J. G. (1983) 'Acid rain and soil microbial activity: effects and their mechanisms', *Journal of Environmental Quality* 12: 133–7.

Kim, Y. S. (1985) 'Air pollution, climate, socioeconomic status and total mortality in the United States', *Science of the Total Environment* 42: 1245–56.

Kinniburgh, D. G. (1986) 'Towards more detailed methods for quantifying the acid susceptibility of rocks and soils', *Journal of the Geological Society* 143: 679–90.

Koide, M. and Goldberg, E. D. (1971) 'Atmospheric sulphur and fossil fuel consumption', *Journal of Geophysical Research* 76: 6589–96.

Krug, E. C. and Frink, C. R. (1983) 'Acid rain in acid soil – a new perspective', *Science* 221: 520–5.

Krug, E. C., Isaacson, P. J., and Frink, C. R. (1985) 'Appraisal of some current hypotheses describing acidification of watersheds', *Journal of the Air Pollution Control Association* 35: 109–14.

Krupta, S. V. and Legge, A. H. (eds) (1982) *Proceedings of an International Conference on Air Pollution Effects on Terrestrial Ecosystems (Banff, 1980)*, New York: Wiley.

Kucera, V. (1976) 'Effects of sulphur dioxide and acid precipitation on metals and anti-rust painted steel', *Ambio* 5: 243–8.

La Bastille, A. (1981) 'Acid rain – how great a menace?' *National Geographic* 160: 652–79.

Last, F. T. (1982) 'Effects of atmospheric sulphur compounds on natural and man-made terrestrial and aquatic ecosystems', *Agriculture and Environment* 7: 299–387.

Last, F. T. and Nicholson, I. A. (1982) 'Acid rain', *Biologist* 29: 250–2.

Lave, L. B. and Seskin, E. P. (1970) 'Air pollution and human health', *Science* 169: 723–33.

Laxen, D. P. H. (1984) 'Linear scale for acid rain?' *Nature* 309: 409.

Lean, G. (1986) 'Acid rain', *The Observer Colour Supplement* 19 October: 50–60.

Lefohn, A. S. and Brocksen, R. W. (1984) 'Acid rain effects research: a status report', *Journal of the Air Pollution Control Association* 34: 1005–13.

Leivestad, H., Hendrey, G., Muniz, I. P., and Snekvik, E. (1976), in F. H. Braekke (ed.) *Impact of Acid Precipitation on Forest and Freshwater Ecosystems in Norway*, Norway: SNSF Project Research Report FR6/76.

Leivestad, H. and Muniz, I. P. (1976) 'Fish kill at low pH in a Norwegian river', *Nature* 259: 391–2.

Lesinski, J. A. (1983) 'Effects of acid precipitation on forests in Poland: selected topics', *Water Quality Bulletin* 8: 156–9, 169.

Likens, G. E. (1976) 'Acid precipitation', *Chemical and Engineering News* 54: 29–37.

Likens, G. E. (1985) 'An experimental approach for the study of ecosystems', *Journal of Ecology* 73: 381–96.

Likens, G. E. and Bormann, F. H. (1974) 'Acid rain – a serious regional environmental problem', *Science* 184: 1176–9.

Likens, G. E. and Butler, T. J. (1981) 'Recent acidification of precipitation in North America', *Atmospheric Environment* 15: 1103–9.

Likens, G. E., Bormann, F. H., and Johnson, N. M. (1972) 'Acid rain', *Environment* 14: 33–40.

Likens, G. E., Johnson, N. M., Galloway, J. N., and Bormann, F. H. (1976) 'Acid precipitation: strong and weak acids', *Science* 194: 643–5.

Likens, G. E., Bormann, F. H., Pierce, R. S., Eaton, J. S., and Johnson, N. M. (1977) *Biogeochemistry of a Forested Ecosystem*, New York: Springer-Verlag.

Likens, G. E., Wright, R. F., Galloway, J. N., and Butler, T. J. (1979) 'Acid rain', *Scientific American* 241: 43–51.

Liljelund, L. E. (1985) 'Nature endangered', *Acid Magazine* 3: 4–5.

Lines, R. (1984) 'Species and seed origin trials in the industrial Pennines', *Quarterly Journal of Forestry* 78: 9–23.

Linthurst, R. A. (ed.) (1984) *Direct and Indirect Effects of Acid Deposition on Vegetation*, London: Butterworth.

Lipfert, F. W. (1985) 'Mortality and air pollution: is there a meaningful connection?' *Environmental Science and Technology* 19: 764–70.

Ludwig, J. H. (1969) 'Air pollution control technology: research and development on new and improved systems', pp. 21–30 in C. C. Havighurst (ed.) *Air Pollution Control*, New York: Oceana.

Lunde, G. (1976) 'Long range aerial transmission of organic micropollutants', *Ambio* 5: 207–8.

Lynn, D. (1976) *Air Pollution: Threat and Response*, Reading, MA: Addison Wesley.

Mabey, R. (1974) *The Pollution Handbook*, Harmondsworth: Penguin.

McCarl, B., Raphael, D., and Stafford, E. (1975) 'The impact of man on the world nitrogen cycle', *Journal of Environmental Management* 3: 7–19.

McCormick, J. (1985) *Acid Earth – The Global Threat of Acid Pollution*, London: Earthscan.

McDonald, M. E. (1985) 'Acid deposition and drinking water', *Environmental Science and Technology* 19: 772–6.

McLaughlin, S. B. (1985) 'Effects of air pollution on forests: a critical review', *Journal of the Air Pollution Control Association* 35: 512–34.

Mainzer, U. (1985) 'Atmospheric pollution', *Future for Our Planet* 25: 5–7.

Malmer, N. (1976) 'Acid precipitation: chemical changes in the soil', *Ambio* 5: 231–4.

Mandelbaum, P. (ed.) (1985) *Acid Rain: Economic Assessment*, New York: Plenum.

Mansfield, T. (ed.) (1976) *Effects of Air Pollutants on Plants*, Cambridge: Cambridge University Press.

Manson, A. (1985) 'Acid rain policy – the Canadian perspective', *Journal of the Air Pollution Control Association* 35: 205–9.

Marcus, M. (1986) 'Environmental policies in the United States', pp. 45–76 in C. C. Park (ed.) *Environmental Policies – An International Review*, London: Croom Helm, pp. 45–76.

Marnot-Houdayer, J. (1985) 'Synergetic effects of pollution', *Forum* (Council of Europe) 3: viii–x.

Marsh, A. (1978) 'Sulphur and nitrogen contributions to the acidity of rain', *Atmospheric Environment* 12: 401–6.

Martin, A. (1979) 'A survey of the acidity of rainwater over large areas of Great Britain', *Science of the Total Environment* 13: 119–30.

Martin, A. and Barber, F. (1978) 'Some observations of acidity and sulphur in rainwater from rural sites in central England and Wales', *Atmospheric Environment* 12: 1481–7.

Mason, A. (1978) 'Sulphur and nitrogen contributions to the acidity of rain', *Atmospheric Environment* 12: 1500–14.

Mason, B. J. (1984) *The Current Status of Research on Acidification of Surface Waters*, London: Royal Society.

Mason, J. and Seip, H. M. (1985) 'The current status of knowledge on acidification of surface waters and guidelines for further research', *Ambio* 14: 45–51.

Mathews, R. O. *et al.* (1981) 'Acid rain in Ireland', *Irish Journal of Environmental Science* 1: 47–50.

Maurin, P. G. and Jonakin, J. (1970) 'Removing sulphur oxides from stacks', *Chemical Engineering* 77: 173–80.

May, J. (1982) 'Dragon's revenge', *Undercurrents* 54: 9–11.

Meetham, A. R. (1964) *Atmospheric Pollution: Its Origins and Prevention*, London: Pergamon.

Meetham, A. R., Bottom, D. W., Henderson-Sellers, A., Cayton, S., and Chambers, D. (1981) *Atmospheric Pollution: Its History, Origins and Prevention*, London: Pergamon.

Mellanby, K. (1972) *The Biology of Pollution*, London: Arnold.

Mellanby, K. (1984a) 'Acid comments on a cause célèbre', *New Scientist* 1429: 37.

Mellanby, K. (1984b) 'What we know about acid rain', *Nature* 307: 87.

Miller, H. G. and Miller, J. D. (1983) 'The interaction of acid precipitation and forest vegetation in northern Britain', *Water Quality Bulletin* 8: 121–6.

Millett, E. J. (1978) 'Chemicals: how dangerous are they?' *Engineering* 218: 1014–16.

Moran, J. M., Morgan, M. D., and Wiersma, J. H. (1980) *Introduction to Environmental Science*, San Francisco: Freeman.

Morgan, M. G. and Morris, M. S. (1984) 'Technical uncertainty in quantitative policy analysis: a sulfur air pollution example', *Risk Analysis* 4: 201–16.

Mosello, R. and Tartari, G. (1983) 'Effects of acid precipitation on subalpine and alpine lakes', *Water Quality Bulletin* 8: 96–100.

Moss, M. R. (1975a) 'Spatial patterns of sulphur accumulation by vegetation and soils around industrial centres', *Journal of Biogeography* 2: 205–22.

Moss, M. R. (1975b) 'Spatial patterns of precipitation reaction', *Environmental Pollution* 8: 301–15.

Mrose, H. (1966) 'Measurement of pH, and chemical analyses of rain, snow and fog waters', *Tellus* 18: 266–70.

Muniz, I. P. (1983) 'The effects of acidification on Norwegian freshwater ecosystems', pp. 299–322 in *Ecological Effects of Acid Deposition*, National Swedish Environment Protection Board Report PM1636.

Muniz, I. P. and Leivestad, H. (1980a) 'Acidification effects on freshwater fish', pp. 84–92 in D. Drablos and A. Tollan (eds) *Ecological Impact of Acid Precipitation*, Norway: SNSF Project.

Muniz, I. P. and Leivestad, (1980b) 'Toxic effects of aluminium on the brown trout (*Salmo trutta* L), pp. 320–1 in D. Drablos and A. Tollan (eds) *Ecological Impact of Acid Precipitation*, Norway: SNSF Project.

Murozumi, M., Chow, T. J., and Patterson, C. (1969) 'Chemical concentrations of pollutant lead aerosols, terrestrial dusts and sea salt in Greenland and Antarctic snow strata', *Geochimica et Cosmochimica Acta* 33: 1247–94.

Myers, N. (ed.) (1985) *The Gaia Atlas of Planet Management* London: Pan.

National Academy of Science (1983) *Acid Deposition – Atmospheric Processes in Eastern North America*, Washington, DC: National Academy Press.

Netter, T. W. (1985) 'Pollution threatening Swiss and German forests', *Environmental Conservation* 12: 27–33.

Nicholson, I. A. and Paterson, I. S. (1983) 'Aspects of acid precipitation in relation to vegetation in the United Kingdom', *Water Quality Bulletin* 8: 59–66.

Nicholson, I. A., Paterson, I. S., and Last, F. T. (1980) *Acid Deposition in Forest Ecosystems*, Cambridge: Institute of Terrestrial Ecology.

Nicholson, I. A., Cape, N., Fowler, D., Kinnaird, J. W., and Patterson, I. S. (1980) 'Effects of a Scots pine (*Pinus sylvestris* L) canopy on the chemical composition and decomposition pattern of precipitation', pp. 148–9 in D. Drablos and A. Tollan (eds) *Ecological Impacts of Acid Precipitation*, Oslo: SNSF.

Nihlgard, B. (1970) 'Precipitation, its chemical composition and effects on soil water in a beech and spruce forest in south Sweden', *Oikos* 21: 208–17.

Nihlgard, B. (1985) 'The ammonium hypothesis: an additional explanation to the forest dieback in Europe', *Ambio* 14: 2–8.

Nilsson, S. I., and Bergvist, B. (1983) 'Aluminium chemistry and acidification processes in a shallow podsol on the Swedish west coast', *Water, Air and Soil Pollution* 20: 311–29.

Nilsson, S. I., Miller, H. G., and Miller, J. D. (1982) 'Forest growth as a possible cause of soil and water acidification: an examination of the concept', *Oikos* 39: 40–9.

Nordo, J. (1976) 'Long range transport of air pollutants in Europe and acid deposition in Norway', *Water, Air and Soil Pollution* 6: 199–217.

Nourse, H. O. (1967) 'The effects of air pollution on house values', *Land Economics* 43: 181–9.

Nriagu, J. O. (ed.) (1980) *Cadmium in the Environment*, New York: Wiley.

Oden, S. (1976) 'The acidity problem – an outline of concepts', *Water, Air and Soil Pollution* 6: 137–66.

Office of Technology Assessment (1984) *Acid Rain and Transported Air Pollutants: Implications for Public Policy*, Washington, DC: US Congress.

Oglesby, A. (1981) 'Acid rain and conservation', *Shooting Times* 19–25 March.

Oglesby, S. (1971) 'Electrostatic precipitators tackle air pollutants', *Environmental Science and Technology* 5: 766–70.

Okita, T. (1983) 'Acid precipitation and related phenomena in Japan', *Water Quality Bulletin* 8: 101–8.

Onnen, J. H. (1972) 'Wet scrubbers tackle pollution', *Environmental Science and Technology* 6: 994–8.

Organization for Economic Co-operation and Development (1977) *Programme on Long Range Transport of Air Pollutants*, Paris: OECD.

Organization for Economic Co-operation and Development (1979) *The OECD Programme on Long Range Transport of Air Pollutants: Measurements and Findings*, Second edition, Paris: OECD.

Organization for Economic Co-operation and Development (1981) *The Costs and Benefits of Sulphur Oxide Control: A Methodological Study*, Paris: OECD.

Organization for Economic Co-operation and Development (1983a) *Control Technology for Nitrogen Oxide Emissions from Stationary Sources*, Paris: OECD.

Organization for Economic Co-operation and Development (1983b) *Costs of Coal Pollution Abatement*, Paris: OECD.

Organization for Economic Co-operation and Development (1985) 'Environment: resources for the future', *OECD Observer* 135: 3–26.

O'Riordan, T. (1986) 'Review of "Acid Earth" by John McCormick', *Ecos* 7: 61–2.

Ormerod, R. (1983) 'Cleaning up Parliament', *Building* 21 October: 28–30.

O'Sullivan, D. A. (1985) 'European concern about acid rain is growing', *Chemical and Engineering News* 63: 12–18.

Ottar, B. (1976) 'Monitoring long range transport of air pollutants: the OECD study', *Ambio* 5: 203–6.

Overrein, L. N., Seip, H. M., and Tollan, A. (1981) *Acid Precipitation – Effects on Forests and Fish*. Final report of the SNSF Project 1972–80, Oslo: SNSF Project.

Paces, T. (1985) 'Sources of acidification in Central Europe, estimate from elemental budgets in small basins, *Nature* 315: 31–6.

Paerl, H. W. (1985) 'Enhancement of primary productivity by nitrogen enriched acid rain', *Nature* 315: 747–9.

Page, J. (1985) 'Stinging in the rain', *Laboratory Practice* November: 33–4.

Park, C. C. (1981) *Ecology and Environmental Management*, London: Butterworth.

Park, C. C. (1983) *Environmental Hazards*, London: Macmillan.

Park, C. C. (ed.) (1986) *Environmental Policies – An International Review*, London: Croom Helm.

Paterson, T. (1984) *A Role for Britain in the Acid Rain Storm*, London: Bow Publications.

Patrinos, A. A. N. (1985) 'The impact of urban and industrial emissions on mesoscale precipitation quality', *Journal of the Air Pollution Control Association* 35: 719–27.

Paulsson, V. (1984) 'Acidification – no new problem issue in Sweden', *Acid Magazine* 1: 4–5.

Pawlick, T. (1984) *A Killing Rain*, New York: Sierra Club Books.

Pawlick, T. (1985) 'What's killing Canada's sugar maples?' *International Wildlife* 5: 34–40.

Pearce, F. (1982a) 'The menace of acid rain', *New Scientist* 95: 419–24.

Pearce, F. (1982b) 'Science and politics don't mix at acid rain debate', *New Scientist* 95: 3.

Pearce, F. (1982c) 'It's an acid wind that blows nobody any good', *New Scientist* 95: 80.

Pearce, F. (1985) 'CEGB is left out in the rain', *New Scientist* 1445: 10–11.

Pearce, F. (1987) *Acid Rain*, Harmondsworth: Penguin.

Penkett, S. A. (1984) 'Ozone increases in ground level European air', *Nature* 311: 14–15.

Pennington, W. (1984) 'Long term natural acidification of upland sites in Cumbria: evidence from post-glacial lake sediments', *Reports of the Freshwater Biological Association* 52: 28–46.

Perhac, R. M. (1985) 'Acid deposition: state of the science', *Journal of the Air Polution Control Association* 35: 198–200.

Perkins, H. C. (1974) *Air Pollution*, New York: McGraw-Hill.

Porritt, J. (1984) *Seeing Green*, London: Blackwell.

Postel, S. (1984) *Air Pollution, Acid Rain and the Future of Forests*, Washington, DC: Worldwatch Institute.

Priest, J. (1973) *Problems of Our Physical Environment*, New York: Addison Wesley.

Rasool, S. I. (ed.) (1973) *Chemistry of the Lower Atmosphere*, New York: Plenum.

Reed, L. E. (1976) 'The long range transport of air pollutants', *Ambio* 5: 202.

Rees, J. (1977) 'The economics of environmental management', *Geography* 62: 311–44.

Rhodes, S. L. (1984) 'Superfunding acid rain controls – who will bear the costs?' *Environment* July/August: 25–31.

Rhodes, S. L. and Middleton, J. (1983) 'The complex challenge of controlling acid rain', *Environment* April: 6–10.

Ribault, J. P. (1984) 'Policies at odds, not researchers', *Forum* (Council of Europe) 4: 8.

Ridley, M. (1985) 'Acid rain: the risk to our stone buildings', *Estates Gazette* 274: 1320–2.

Roberts, T. M. (1984) 'Long term effects of sulphur dioxide on crops: an analysis of dose–response relations', *Proceedings of the Royal Society* 305B: 299–316.

Roberts, T. M. *et al.* (1983) 'Effects of gaseous air pollutants on agriculture and forestry in the United Kingdom', *Advances in Applied Biology* 9: 1–142.

Robson, A. (1979) 'Environmental implications of fossil-fuelled power stations', *Journal of the Royal Society of Health* 99: 247–58.

Robson, A. J. (1977) 'The effect of urban structure on the concentration of pollution', *Urban Studies* 14: 89–93.

Rodhe, H. and Grant, L. (1984) 'An evaluation of sulphate in European precipitation, 1955–1982', *Atmospheric Environment* 18: 2627–39.

Rodhe, H., Crutzen, P., and Vanderpool, A. (1981) 'Formation of sulphuric and nitric acid in the atmosphere during long-range transport', *Tellus* 33: 132–41.

Roope, G. W. (1985) 'Acid rain legislation – a chance for effective public policy', *Journal of the Air Pollution Control Association* 35: 211–13.

Rose, C. (1985) 'Acid rain falls on British woodlands', *New Scientist* 108: 52–3.

Rosenkranz, A. (1983) 'The Stockholm conference, 1982', *Acid News* 1: 4–8.

Royal College of Physicians (1970) *Air Pollution and Health*, London: Pitman.

Royal Commission on Environmental Pollution (1984) *Tenth Report*, London: HMSO.

Royal Ministry of Foreign Affairs (1971) *Air Pollution across National Boundaries: The Impact on the Environment of Sulphur in Air and Precipitation*, Stockholm: Norsted & Soner.

Royal Society (1984) *The Current Status of Research on Acidification of Surface Waters*, London: Royal Society.

Rubin, E. S. (1983) 'International pollution control costs of coal fired power stations', *Environmental Science and Technology* 17: 366–77.

Sawyer, S. W. (1986) *Renewable Energy: Progress, Prospects*, Washington, DC: Association of American Geographers.

Schindler, D. W., Mills, K. H., Malley, D. F., Findlay, D. L., Shearer, J. A., Davis, I. J., Turner, M. A., Linsey, G. A., and Cruickshank, D. R. (1985) 'Long term ecosystem stress: the effects of years of experimental acidification on a small lake', *Science* 228: 1395–401.

Schnoor, J. L. (ed.) (1984) *Modelling of Total Acid Precipitation Impacts*, London: Butterworth.

Schofield, C. L. (1976) 'Acid precipitation: its effects on fish', *Ambio* 5: 228–30.

Schrenk, H. H., Heimann, J., Clayton, G. D., Gafefer, W. M., and Wexler, H. (1949) 'Air pollution in Donora, Pennsylvania: epidemiology of the unusual smog episode of October 1948', *Public Health Bulletin* 306.

Schroeder, T. and Grant, L. (eds) (1982) *Air Pollution by Nitrogen Oxides*, Amsterdam: Elsevier.

Schroter, H. (1985) 'Is the West German experience of acid rain different?' pp. 120–34 in Scottish Wildlife Trust, *Report of the Acid Rain Inquiry, 27–29*

September 1984, Edinburgh: Scottish Wildlife Trust.

Scott, J. A. (1963) 'The London fog of December 1962', *The Medical Officer* 109: 250.

Scottish Wildlife Trust (1985) *Report of the Acid Rain Inquiry, 27–29 September 1984*, Edinburgh: Scottish Wildlife Trust.

Scriven, R. (1985) 'What are the sources of acid rain?' pp. 13–27 in Scottish Wildlife Trust, *Report of the Acid Rain Inquiry, 27–29 September 1984*, Edinburgh: Scottish Wildlife Trust.

Serrill, M. S. (1986) 'Anatomy of a catastrophe', *Time* 35: 6–11.

Sevaldrup, I., Muniz, I. P., and Kalvenes, S. (1980) 'Loss of fish populations in southern Norway: dynamics and magnitude of the problem', pp. 350–1, in D. Drablos and A. Tollan (eds) *Ecological Impacts of Acid Precipitation*, Oslo: SNSF.

Sharp, A. D., Trudgill, S. T., Cooke, R. U., Price, C. A., Crabtree, R. W., Pickles, A. M., and Smith, D. I. (1982) 'Weathering of the balustrade on St Paul's Cathedral, London', *Earth Surface Processes and Landforms* 7: 387–9.

Skeffington, R. A. and Roberts, T. M. (1985) 'Effects of ozone and acid mists on Scots pine saplings', *Oecologia* 65: 201–6.

Slack, A. V. (1973) 'Removing SO_2 from stack gases', *Environmental Science and Technology* 7: 110–19.

Smith, D. B. (1980) 'The effects of afforestation on the trout of a small stream in southern Scotland', *Fisheries Management* 11: 39–58.

Smith, R. A. (1852) 'On the air and rain of Manchester', *Memoirs and Proceedings of the Manchester Literary and Philosophical Society* 2: 207–17.

Smith, R. A. (1872) *Air and Rain: The Beginnings of a Chemical Climatology*, London: Longmans Green.

Spence, K. (1985) 'Saving the queen of spires', *Country Life* 177: 729–31.

Stengel, R. (1986) 'The Lady's party', *Time* 28: 6–12.

Stern, A. C. (ed.) (1968/9) *Air Pollution*, 3 volumes, London: Academic Press.

Stern, A. C., Wohlers, H. C., Bonbel, R. W., and Lowry, W. P. (1973) *Fundamentals of Air Pollution*, New York: Academic Press.

Stevenson, C. M. (1968) 'An analysis of the chemical composition of rainwater and air over the British Isles and Eire for the years 1959–64', *Quarterly Journal of the Royal Meteorological Society* 94: 56–70.

Stoner, J. H., Gee, A. S., and Wade, K. R. (1984) 'The effects of acidification on the ecology of streams in the upper Tywi catchment in west Wales', *Environmental Pollution* 35: 125–57.

Straud, L. (1980) 'The effect of acid precipitation on tree growth', pp. 64–7 in D. Drablos and A. Tollan (eds) *Ecological Impacts of Acid Precipitation*, Oslo: SNSF Project.

Strauss, W. and Mainwaring, S. J. (1984) *Air Pollution*, London: Arnold.

Streets, D. G., Knudson, D. A., and Shannon, J. D. (1983) 'Selected strategies to reduce acid deposition in the United States', *Environmental Science and Technology* 17: 474–85.

Stretton, C. (1984) 'Water supply and forestry – a conflict of interests: Cray Reservoir, a case study', *Journal of the Institute of Water Engineers and Scientists* 38: 323–30.

Summers, P. W. (1983) 'A global perspective on acid deposition, its sources and atmospheric transport', *Water Quality Bulletin* 8: 81–88, 109.

Sutcliffe, D. W. (1983) 'Acid precipitation and its effects on aquatic systems in the English Lake District (Cumbria)', *Reports of the Freshwater Biological Association* 51: 30–62.

Sutcliffe, D. W. and Carrick, T. R. (1986) 'Effects of acid rain on waterbodies in

Cumbria', pp. 16–24 in P. Ineson (ed.) *Pollution in Cumbria*, Grange over Sands: ITE.

Sutcliffe, D. W., Carrick, T. R., Heron, J., Rigg, E., Talling, J. F., Woof, C., and Lund, J. W. G. (1982) 'Long term and seasonal changes in the chemical composition of precipitation and surface waters of lakes and tarns in the English Lake District', *Freshwater Biology* 12: 451–506.

Svensson, G. (1984) 'Forest at risk?' *Acid Magazine* 1: 12–27.

Swedish Ministry of Agriculture (1982) *Acidification Today and Tomorrow*, Stockholm: Swedish Ministry of Agriculture.

Swedish Ministry of Agriculture (1984) *Acidification – A Boundless Threat to Our Environment*, Solna: National Swedish Environment Protection Board.

Swedish Natural Science Research Council (1976) *Nitrogen, Phosphorus and Sulphur – Global Cycles*, SCOPE Report 7, Stockholm.

Swedish Society for the Conservation of Nature (1981) *Acid Rain*, Stockholm: Swedish Society for the Conservation of Nature.

Tabatari, M. A. (1985) 'Effects of acid rain on soils', *Critical Reviews in Environmental Control* 15: 65–110.

Tallis, J. H. (1964) 'Studies on southern Pennine peats III. The behaviour of sphagnum', *Journal of Ecology* 52: 345–53.

Tamm, C. O. and Cowling, E. B. (1977) 'Acid precipitation and forest vegetation', *Water, Air and Soil Pollution* 7: 503–11.

Tamm, C. O., Wiklander, G., and Popovic, B. (1977) 'Effects of application of sulphuric acid to poor pine forest', *Water, Air and Soil Pollution* 8: 75–87.

Taylor, P. (1985) 'Acid rain – Cumbria's unwanted visitor', *Lake District Guardian*, Kendal: Lake District Special Planning Board.

Thompson, R. (1978) *Atmospheric Contamination – A Review of the Air Pollution Problem*, Reading Geographical Papers, University of Reading, England.

Thomson, G. (1969) 'Sulphur dioxide damage to antiquities', *Atmospheric Environment* 3: 687.

Thornton, I. and Plant, J. (1980) 'Regional geochemical mapping and health in the United Kingdom', *Journal of the Geological Society of London* 137: 575–86.

Thornton, I. and Webb, J. S. (1979) 'Geochemistry and health in the United Kingdom', *Philosophical Transactions of the Royal Society of London* 288B: 151–68.

Tifft, S. (1985) 'Requiem for the forest', *Time* 126: 46–53.

Tolba, M. K. (1983) 'Acid rain – a growing concern of industrialized countries', *Water Quality Bulletin* 8: 115–20, 167.

Tollan, A. (1981) 'Effects of acid precipitation on inland waters', *Geo Journal* 5: 409–16.

Tomlinson, G. (1983) 'Air pollutants and forest decline', *Environmental Science and Technology* 17: 246–56.

Toribara, T. Y., Miller, M. W., and Morrow, P. E. (eds) (1980) *Polluted Rain*, New York: Plenum.

Tukey, H. B. (1979) 'The leaching of substances from plants', *Annual Reviews of Plant Physiology* 21: 305–24.

Tunnicliffe, M. (1975) *Industrial Air Pollution*, London: HMSO.

Ulrich, B., Mayer, R., and Khanna, P. K. (1980) 'Chemical changes due to acid precipitation in a loess-derived soil in central Europe', *Soil Science* 130: 193–9.

United Kingdom Review Group on Acid Rain (1982) *Acidity of Rainfall in UK – A Preliminary Report*, Stevenage: Warren Spring Laboratory.

United Kingdom Review Group on Acid Rain (1984) *Acid Deposition in the United*

Kingdom, Stevenage: Warren Spring Laboratory.

United Nations Economic Commission for Europe (1982) *Effects of Sulphur Compounds on Materials, Including Historic and Cultural Monuments*, Geneva: UNECE.

United Nations Economic Commission for Europe (1984) *Airborne Sulphur Pollution: Effects and Controls*, New York: UNECE.

United States–Canada Memorandum of Intent on Transboundary Air Pollution (1981) *Impact Assessment*, Phase II interim working paper, Toronto.

United States Environmental Protection Agency (1982) *Sulfur Emissions: Control Technology and Waste Management*, Washington, DC: Office of Research and Development.

United States Environmental Protection Agency (1983) *The Acidic Deposition Phenomenon and its Effects – Critical Assessment Review Papers*, Washington, DC: Environmental Protection Agency.

Unsworth, M. H. (1984) 'Evaporation from forests in cloud enhances the effects of acid deposition, *Nature* 312: 262–4.

Van Breeman, N., Burrough, P. A., Velthorst, E. J., Van Dobben, H. F., de Wit, T., Riddler, T. B., and Reijnders, H. F. R. (1982) 'Soil acidification from atmospheric ammonium sulphate in forest canopy throughfall', *Nature* 299: 548–50.

Van Breeman, N., Mulder, J., and Driscoll, C. T. (1983) 'Acidification and alkalinization of soils', *Plant and Soil* 75: 283–308.

Van Breeman, N., Driscoll, C. T., and Mulder, J. (1984) 'Acidic deposition and internal proton sources in acidification of soils and waters', *Nature* 307: 599–604.

Van Cleef, E. (1908) 'Is there a type of storm path?' *Monthly Weather Review* 36: 56–8.

Vangenechten, J. H. D. (1983) 'Acidification in west European lakes and hydrobiological adaptation to acid stress in natural inhabitants of acid lakes', *Water Quality Bulletin* 8: 150–5, 169.

Vogelman, W. H. (1982) 'Catastrophe on Camel's Hump', *Natural History* November: 8–14.

Waggoner, A. P., Vanderpool, A. J., Charlson, R. J., Larsen, S., Granat, L., and Trägårdh, C. (1976) 'Sulphate light scattering ratio as an index of the role of sulphur in tropospheric optics', *Nature* 261: 120–2.

Walgate, R. (1984) 'CEGB takes a pasting from MPs', *Nature* 311: 94.

Walling, D. E. and Webb, B. W. (1981) 'Water quality', pp. 126–72 in J. Lewin (ed.) *British Rivers*, London: Allen & Unwin.

Ward, B. and Dubos, R. (1972) *Only One Earth*, Harmondsworth: Penguin.

Watt Committee on Energy (1984) 'Acid rain', papers presented at the 15th Consultative Council Meeting of the Watt Committee on Energy, London, 1 December 1983.

Wetstone, G. S. and Rosenkranz, A. (1983) *Acid Rain in Europe and North America – National Responses to an International Problem*, Washington, DC: Environmental Law Institute.

Whelpdale, D. M. (1983) 'Acid deposition: distribution and impact', *Water Quality Bulletin* 8: 72–80.

White, D. (1979) 'How polluted are we, and how polluted do we want to be?' *New Society* 47: 63–6.

Wiklander, L. and Anderson, A. (1972) 'The replacing efficiency of hydrogen ion in relation to base saturation and pH', *Geoderma* 7: 159–65.

Wilkins, E. T. (1954) 'Air pollution and the London fog of December 1952', *Journal of the Royal Sanitary Institute* 74: 1–21.

Winkler, E. M. (1970) 'The importance of air pollution in the corrosion of stone and metals', *Enginering Geology* 4: 327–34.

Wise, W. (1971) *Killer Smog*, New York: Anderson-Ballantine.

Wolff, E. (1986) 'Climate, pollution and ice', *NERC Newsjournal* 9: 4–7.

Wolff, E. and Peel, D. (1985) 'The record of global pollution in polar snow and ice', *Nature* 313: 535–40.

Wood, C. and Lawrence, M. (1980) 'Air pollution and human health in Greater Manchester', *Environment and Planning* 12A: 1427–39.

Woodwell, G. M. (1970) 'Effects of pollution on the structure and physiology of ecosystems', *Science* 168: 429–33.

World Health Organisation (1984) *Urban Air Pollution, 1973–80*, Geneva: WHO.

Wright, R. F. (1977) *Historical Changes in the pH of 128 Lakes in Southern Norway and 130 Lakes in Southern Sweden over the Years 1923–1976*, Oslo: SNSF Project TN34/77.

Wright, R. F. (1983) 'Acidification of freshwaters in Europe', *Water Quality Bulletin* 8: 137–42, 168.

Wright, R. F. and Gjessing, E. T. (1976) 'Acid precipitation: changes in the chemical composition of lakes', *Ambio* 5: 219–23.

Wright, R. F. and Henriksen, A. (1978) 'Chemistry of small Norwegian lakes, with special reference to acid precipitation', *Limnology and Oceanography* 23: 487–98.

Wright, R. F. and Henriksen, A. (1983) 'Restoration of Norwegian lakes by reduction in sulphur deposition', *Nature* 305: 422–4.

Wright, R. F., Conroy, N., Dickson, W. T., Harriman, R., Henriksen, A., and Schofield, C. L. (1980) 'Acidified lake districts of the world', pp. 377–9 in D. Drablos and A. Tollan (eds) *Ecological Impacts of Acid Precipitation*, Oslo: SNSF.

Wright, R. F., Harriman, R., Henriksen, A., Morrison, B., and Caires, L. A. (1980) 'Acid lakes and streams in the Galloway area, southwestern Scotland', pp. 248–9 in D. Drablos and A. Tollan (eds) *Ecological Impacts of Acid Precipitation*, Oslo: SNSF.

Yocom, J. E. and Upham, J. B. (1977) 'Effects on economic materials and structures', pp. 65–115 in A. C. Stern (ed.) *Air Pollution*, volume 3, New York: Academic Press.

INDEX